HYPERSKEW POLYHEDRA

BEING THE NINTH PART
OF SEVERAL COMPRISING
THE COMPLETE? POLYHEDRA
BY PATRICK TAYLOR

HYPERSKEW POLYHEDRA

First published 2017 by

Polystar Press

277 Cavendish Street	62 Angel Street
Ipswich	Hadleigh
Suffolk	Suffolk
IP3 8BQ	IP7 5EY
e-mail: polystar @ntlworld.com	@btinternet.com
(01473) 434604	(01473) 824896

Made in Cornwall

ISBN 978 1 907154 70 6

POLYSTAR PRESS IPSWICH

CONTENTS

ILLUSTRATIONS

Figures * are supported by Tables nos. 1-54

for Annie and Jessica, our brown eyed girls

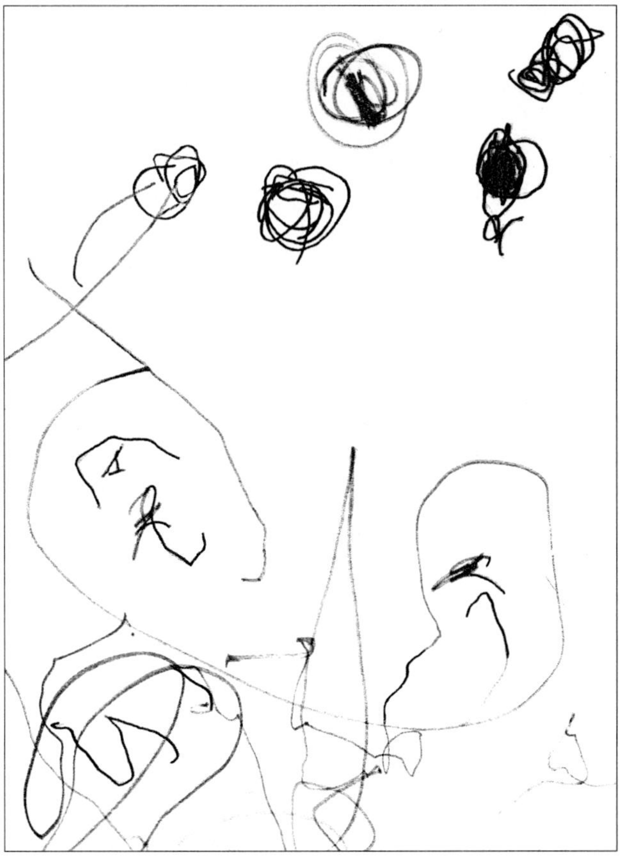

"There is the science of pure geometry, in which there are many geometries, ... each of [which] is a *model*, a pattern of ideas, and is to be judged by the interest and beauty of its particular pattern."

G H Hardy *A Mathematician's Apology* 1940

INTRODUCTION

In the last volume of this series we looked at the skew polyhedra and their degenerate forms, surfaces comprised of regular polygons that wrap around the closepacks and tilings. These were arranged in families where the basic 4-fold, 6-fold, 8-fold etc. connectivities between adjoining cells were preserved, despite the polygons being changed by vertex transformations.

All of these forms repeated themselves to infinity, the skew polyhedra in the three dimensions of Euclidean space, the degenerates in just two dimensions.

We have also seen in previous volumes how the tilings extend to infinity, but can be made to form the polyhedra by putting fewer or smaller valued polygons around a vertex, effectively becoming tilings of a sphere. By analogy the closepacks that fill three dimensions, can be made to fold up into four dimensional closed figures, the hypersolids, by putting fewer or smaller valued polyhedra around an edge.

The techniques employed in the last volume, to wrap surfaces around the closepacks and tilings, producing the skew polyhedra and their degenerates respectively, will now be applied to these folded up four dimensional forms, the hypersolids, to produce what have to be described as hyperskew polyhedra, regular surfaces in four dimensions, a systematic description of which has not been undertaken before.

The final part of this volume will then apply the same techniques to the polyhedra to produce the degenerate hyperskews, simpler forms, only just four dimensional, bringing us full circle back to where we started.

FOLDING VERTICES

We saw in the last volume that the notional second truncation of the 4-connected truncated tetrahedral tetraskew {6,6 |3} has two plus two hexagons at each vertex. We thus found the 6-connected truncated octahedral hexaskew {6,4 |4} with a totally different set of symmetries and a tetrahedral vertex figure comprising four hexagons around two missing squares.

Later in the same volume we encountered amongst the degenerate skew polyhedra, the truncated tritile triskew {6,4 |6}, also with four hexagons at each vertex, but now with a missing hexagon and digon. The vertex here is essentially a folded back version of the previous one and naturally leads to the question of whether any regular form might exist in between, perhaps with a missing pentagon and triangle?

Folding such a vertex back in three dimensions, both a triangle and pentagon are encountered but not at the same time. Such a form can be considered, but is not realisable in just three dimensions, otherwise dodecahedra would be found to closepack. This form's hexagons wrap around the hypersolid t^3{5,3,3}, in the same way those of the skew polyhedron above wrap around the closepack t^3{4,3,4} and those of the degenerate above wrap around the tiling polytope t^3{6,3,2}.

The surface produced comprises all 2,400 hexagons of this hypertruncated icosatetrahedron. Its 600 4-connected truncated tetrahedra are linked by their missing triangular faces, forming a complementary space to its 120 12-connected truncated icosahedra linked by their missing pentagons, each space is one side or the other of any hexagon face. Because of the different missing faces, we need to expand Coxeter's notation to record both in the format {6,4 |5 3}.

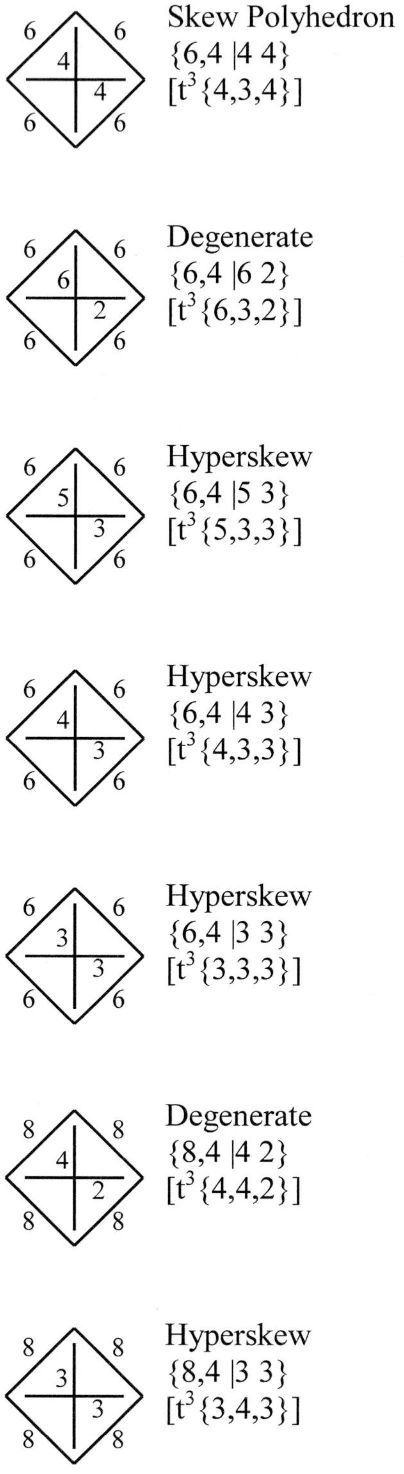

Fig.1: Tetrahedral Vertex Parents

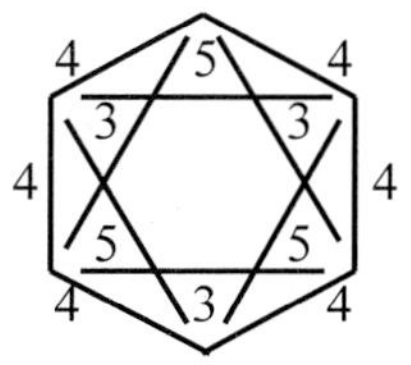

Hyperskew
{4,6 |5 3}
[o^3{5,3,3}]

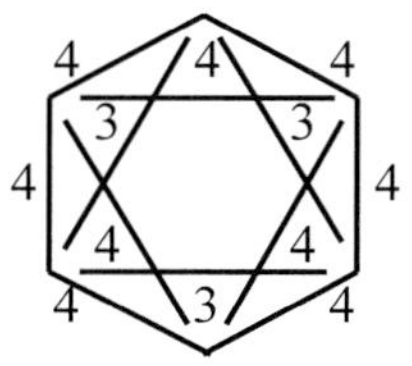

Hyperskew
{4,6 |4 3}
[o^3{4,3,3}]

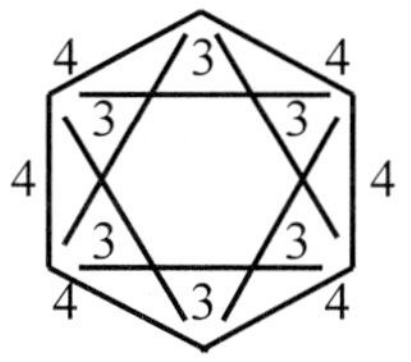

Hyperskew
{4,6 |3 3}
[o^3{3,3,3}]

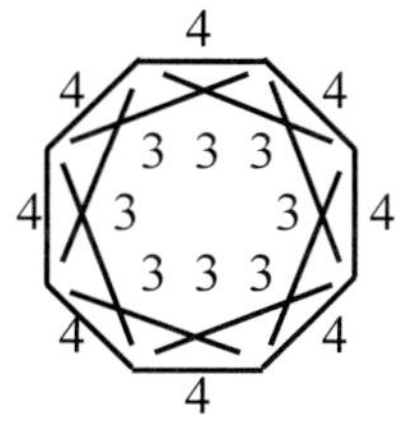

Hyperskew
{4,8 |3 3}
[o^3{3,4,3}]

Fig.2: Dual Hyperskew Parents

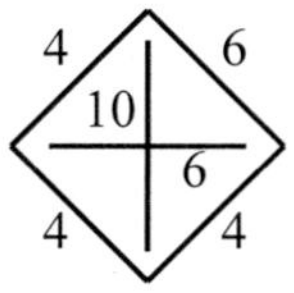

Hyperskew
[o{5,3,3}]

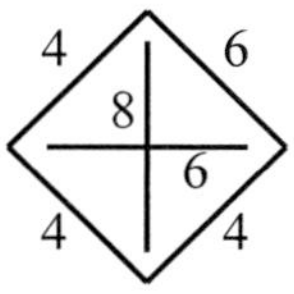

Hyperskew
[o{4,3,3}]

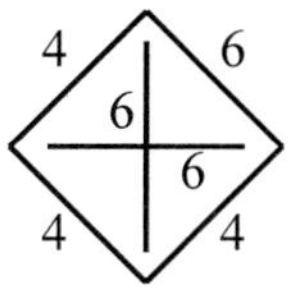

Hyperskew
[o{3,3,3}]

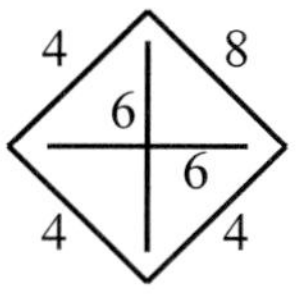

Hyperskew
[o{3,4,3}]

Fig.3: Even Faced Hyperskews

Folding such a vertex up even tighter in the fourth dimension, two further possibilities emerge, one with missing squares and triangles {6,4 |4 3}, its hexagons wrapped around the hypertruncated octetrahedron t^3{4,3,3}, the other with two sets of missing triangles {6,4 |3 3}, its hexagons wrapped around the hypertruncated tetrahedron, t^3{3,3,3}.

Also in the last volume we looked for a figure analogous to the {6,4 |4}, but derived instead as a notional second truncation of the 8-connected truncated cubic octaskew {8,8 |3}. All that we found was a degenerate, the 4-connected truncated tetratile tetraskew {8,4 |4}, which again had a tetrahedral vertex figure with four octagons around a missing square and digon, more correctly {8,4 |4 2}.

This vertex can be folded back, ultimately to itself as {8,4 |2 4}, via another intermediate form which has two missing triangles, but is again not realisable in three dimensions. This produces an 8-connected hyperskew {8,4 |3 3}, comprising all 144 octagons of a hypertruncated cube t^3{3,4,3}. The surface separates two equal sets of 24 truncated cubes each linked by their missing triangle faces.

There is thus a hyperskew polyhedron related to the central 't^3' member of each of the four hypersolid symmetry families. As with the skew polyhedra, each of these will be found to have a dual related to the respective third odd 'o^3' member (Figure 2) of each family and a grid of related forms derived by vertex transformations from the first odd 'o' member (Figure 3) of each family. We will now look at each of these in turn, attempting to derive all possible hyperskew polyhedra.

DODECATETRASKEW GRID

By analogy with the three dimensional skew polyhedra, the parent hyperskews {6,4 |5 3} and {4,6 |5 3}, based on the t^3 and o^3{5,3,3} respectively, have a common 'even faced' stretched version based on the tetrahedral vertex figure of o{5,3,3}. This last has a hexagon and three squares at each vertex with missing decagons and hexagons. The complete figure has two separate spaces: one with 12-connected truncated icosidodecahedra linked by decagon prisms, the other with 4-connected truncated octahedra linked by hexagon prisms.

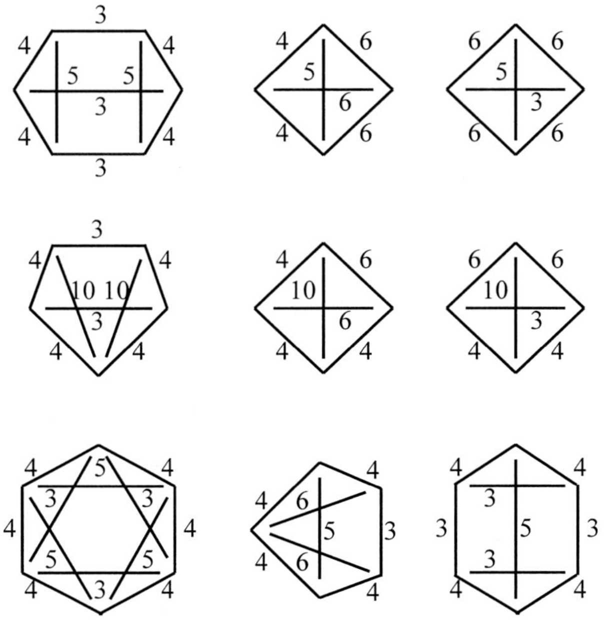

Fig.4: Dodecatetraskew Grid [2-4,2-12]

Figure 4 adjoining shows the full grid for this family, all members of which comprise essentially these same two spaces, one 12-connected the other 4-connected, hence the term 'Dodecatetraskews'. These are all produced by the same process of vertex transformation as was described in *'The Skew Polyhedra',* whereby the central vertex figure is expanded and some of the faces present become 'untruncated'. This produces four 'intermediates' around the central member and four 'resultants' at the corners of the grid. Two of these last are the parent hyperskews {6,4 |5 3}and {4,6 |5 3}.

The two parents and their shared stretched central version are described numerically in Table 1 and shown opposite in Figure 5. Although the numbers seem large, these are finite assemblies of polygons that actually close up in four dimensions.

Figure 6 and Table 2 on page 10 show the bottom right members of this grid, two intermediates based on the tt^2 and o^2{5,3,3} and the resultant in the corner based on t^2t^2{5,3,3}. Similarly Figure 7 and Table 3 on page 11 show the top left members of the grid, two intermediates based on tt^2 and o^2 {3,3,5} and a resultant based on t^2t^2{3,3,5}.

As described in *'Hypersolids and Quasi-Hypersolids'*, the {5,3,3} family of hypersolids are almost closepacks, but do not have quite enough angle around each edge to close without folding up into the fourth dimension.

A two dimensional analogy of a closepack is that of a tiling formed with three hexagons at each vertex, which can fill the plane to infinity. With slightly less angle at each vertex, using two hexagons and a pentagon, it closes by folding up into a third dimension as a truncated icosahedron, a close approximation to a sphere (often used as a pattern on footballs).

For ease of illustration only a small section of each hyperskew will be shown here, as if they did form a closepack. This does involve a slight distortion, which if one tried to draw more than the few cells shown here would become problematic.

DODECATETRASKEWS

Solids	V	E	F	S
tr tetrahedron	3600	7200	2400⬡	600
tr icosahedron			(1200) △	120
{6,4 |5 3}			(720) ⬠	
Hypertruncated Icositetrahedral				
tr octahedron	14400	28800	2400⬡	600
tr icosidodecahedron			10800 □	120
decaprism			(2400) ⬡	720
hexaprism			(1440) ○	1200
nt^2t^2{4,6 |5 3}				
Truncated Hypericosidodecatetraprism				
tetrahedron	2400	7200	3600 □	600
dodecahedron			(2400) △	120
pentaprism			(1440) ⬠	720
triprism				1200
{4,6 |5 3}				
Hyperdodecatetraprism				

Table 1: Dodecatetraskews

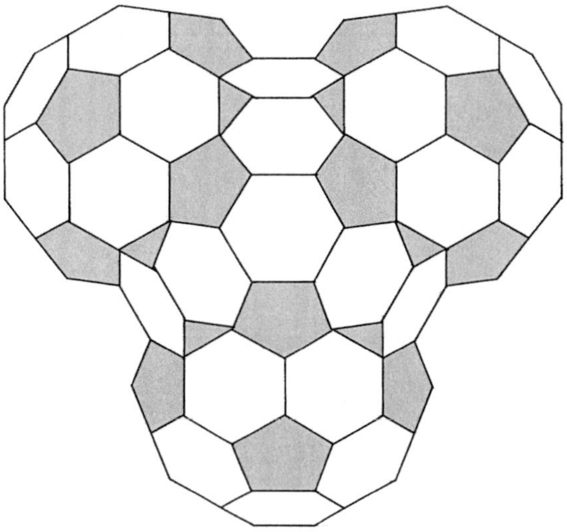

{6,4 |5 3} [t^3{5,3,3}]

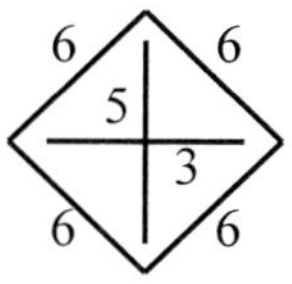

Hypertruncated
Icositetrahedral
Dodecatetraskew

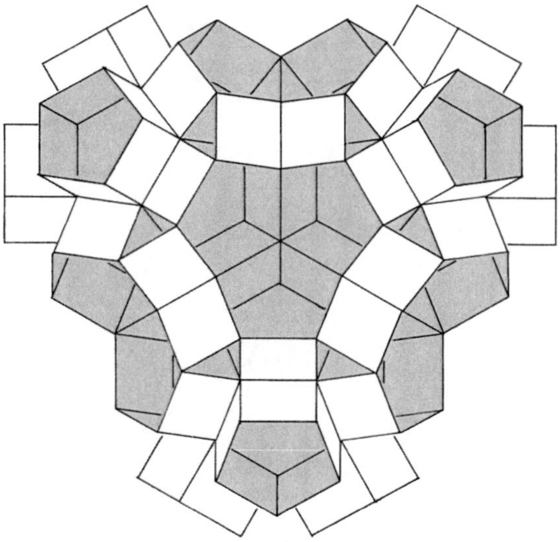

{4,6 |5 3} [o^3{5,3,3}]

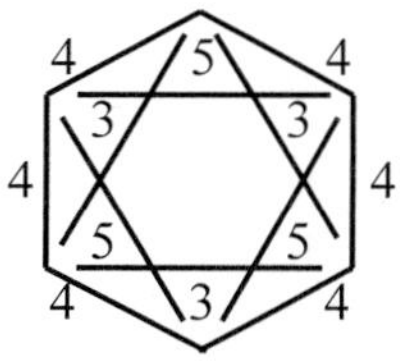

Hyperdodecatetraprism
Dodecatetraskew

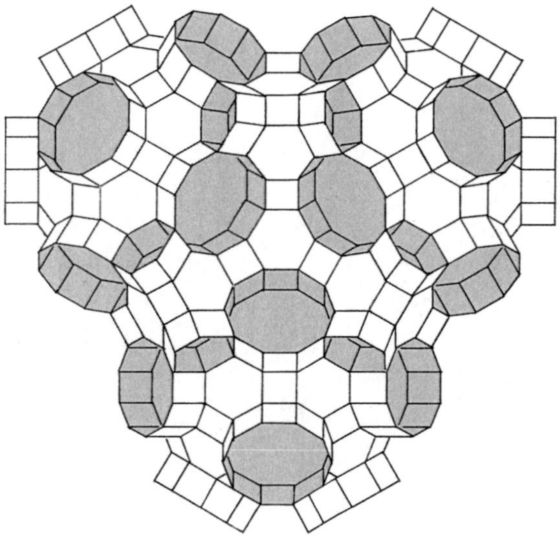

nt^2t^2{4,6 |5 3} [o{5,3,3}]

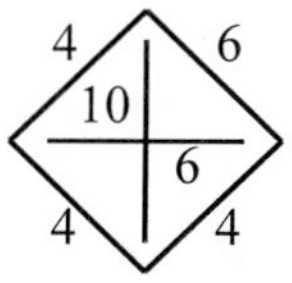

Truncated
Hypericosidodecatetraprism
Dodecatetraskew

Fig.5: Dodecatetraskews

MORE DODECATETRASKEWS

Solids	V	E	F	S
tr tetrahedron	7200	14400	2400⬡	600
tr icosidodecahedron			3600 □	120
triprism			(2400) △	1200
			(720) ◯	
Truncated Hypericosidodecatetrahedral				
octahedron	3600	10800	2400△	600
rhombicosidodecahedron			3600 □	120
triprism			(2400) △	1200
			(720) ⬠	
Hyperhombicosidodecatetrahedral				
tr tetrahedron	7200	18000	2400△	600
rhombicosidodecahedron			7200 □	120
pentaprism			(2400) ⬡	720
hexaprism			(1440) ⬠	1200
Hyperhombicosidodecatetraprism				

Table 2: More Dodecatetraskews

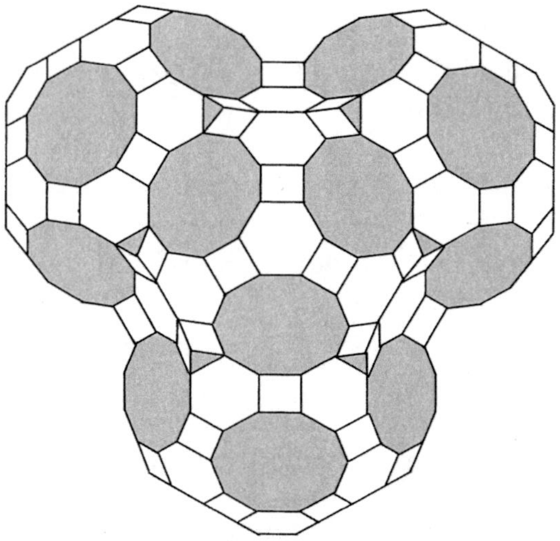

[$tt^2\{5,3,3\}$]
Truncated Hypericosi-
dodecatetrahedral
Dodecatetraskew

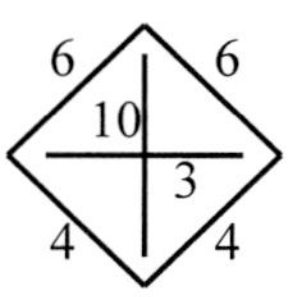

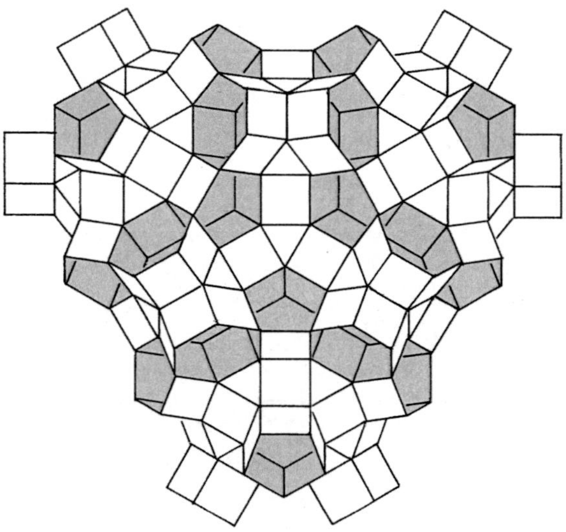

[$o^2\{5,3,3\}$]
Hyperhombicosi-
dodecatetraprism
Dodecatetraskew

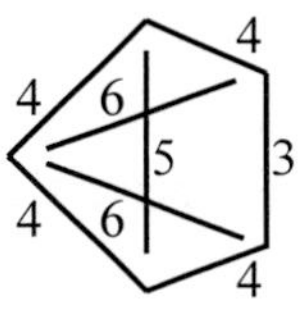

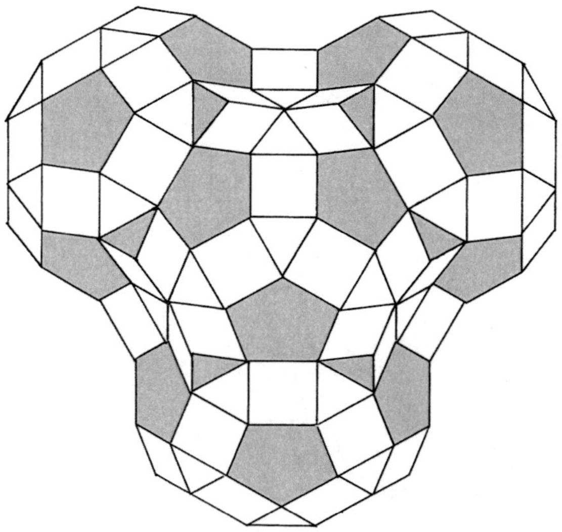

[$t^2t^2\{5,3,3\}$]
Hyperhombicosi-
dodecatetrahedral
Dodecatetraskew

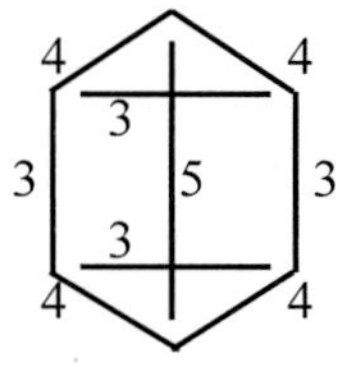

Fig.6: More Dodecatetraskews

EVEN MORE DODECATETRASKEWS

Solids	V	E	F	S
tr octahedron	7200	14400	2400 ⬡	600
tr icosahedron			3600 □	120
pentaprism			(1200) ⬡	720
			(1440) ⬠	
Truncated Hypericositetratetrahedral				
cuboctahedron	3600	10800	2400 △	600
icosidodecahedron			3600 □	120
pentaprism			(1200) △	720
			(1440) ⬠	
Hyperhombicositetratetrahedral				
cuboctahedron	7200	18000	2400 △	600
tr dodecahedron			7200 □	120
decaprism			(2400) △	720
triprism			(1440) ○	1200
Hyperhombicositetratetraprism				

Table 3: Even More Dodecatetraskews

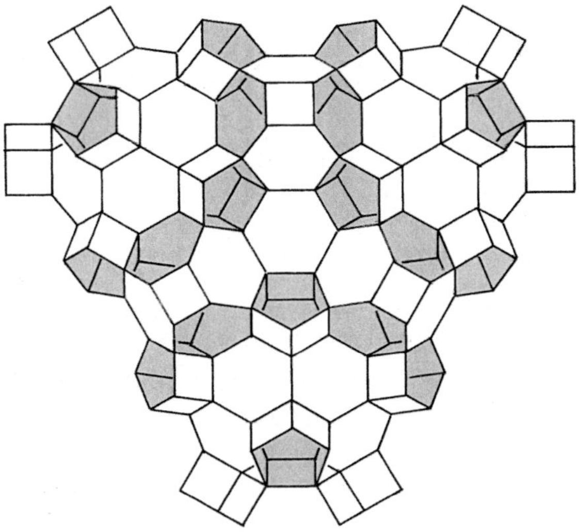

[tt[2]{3,3,5}]
Truncated Hypericositetratetrahedral Dodecatetraskew

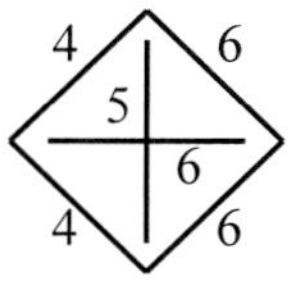

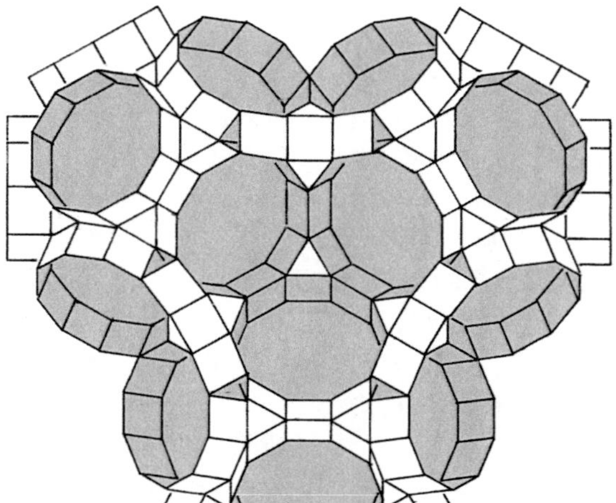

[o[2]{3,3,5}]
Hyperhombicositetratetraprism Dodecatetraskew

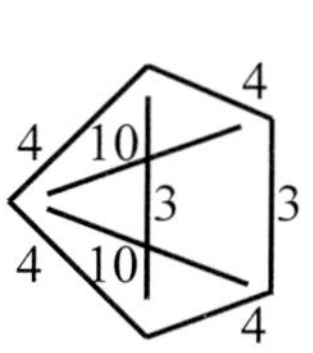

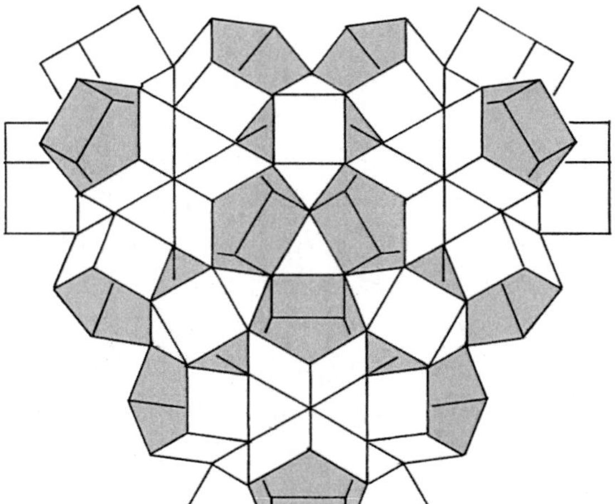

[t[2]t[2]{3,3,5}]
Hyperhombicositetratetrahedral Dodecatetraskew

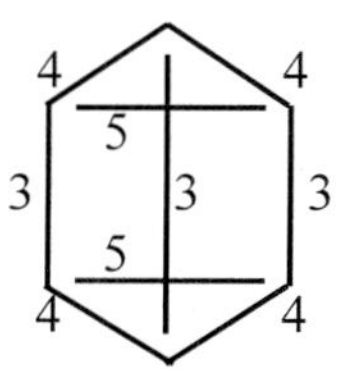

Fig.7: Even More Dodecatetraskews

ICOSAPENTASKEW GRID

The tetrahedral vertex figure of o{5,3,3} can be aligned in three different ways with different paired missing faces to generate three different grids. The first of these was shown in the centre of Figure 4, where the truncated icosidodecahedra were 12-connected, each missing its decagon faces.

The central vertex figure in Figure 8 adjoining shows the same vertex with the truncated icosidodecahedra 20-connected by missing out their hexagon faces. Each of these connects to a 4-connected truncated octahedron to form one space. The other space comprises 3-connected hexagon prisms linked by missing square faces to 5-connected decagon prisms. As with the skew polyhedra, this vertex can be transformed to produce another full grid of nine members as shown in Figure 8. The hyperskews so produced are all basically 20- and 5-connected, hence the term ‘Icosapentaskews’.

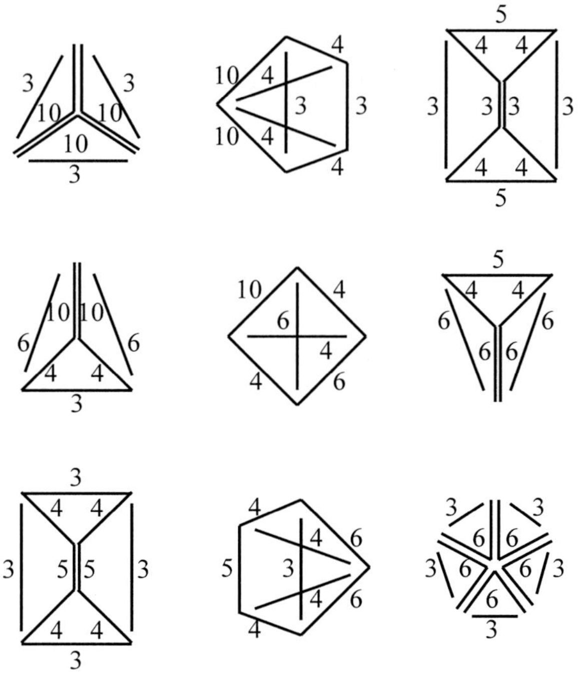

Fig.8: Icosapentaskew Grid [3-5,4-20]

An interesting property of the Dodecatetraskews we have just examined is that their Euler Number, V-E+F, is constant and in fact equal to –1200. This number is 2 for a sphere or any polyhedron, 0 for a torus and takes various other increasingly more negative values as the complexity of a topological surface increases. The fact that it is a constant across all this family is not surprising when it is considered that they are all essentially the same topologically, comprising a single continuous surface between two spaces, one 12-connected the other 4-connected.

The Icosapentaskews that follow will also be seen to have a constant Euler Number, this time –3360, however to achieve it one has to account properly for multiple edges and faces where they occur.

Two of the resultants in this grid can be regarded as a dual pair of parents with amended Coxeter notation {10,6 |3 2} and {6,10 |3 2}, top left and bottom right respectively. These are based on t{5,3,3} and t{3,3,5} respectively, and along with the central member are shown adjoining in Figure 9, with Table 4 describing their numerical quantities.

Figure 10 and Table 5 on page 14 show the bottom left members of this grid, two intermediates based on the tt^2 and o^2{5,3,3} and the resultant in the corner based on t^2t^2{5,3,3}.

Similarly Figure 11 and Table 6 on page 15 show the top right members of the same grid, two intermediates based on tt^2 and o^2 {3,3,5} and a resultant based on t^2t^2{3,3,5}.

ICOSAPENTASKEWS

Solids	V	E	F		S
decagon	2400	3600	1440	⬭	720
tr dodecahedron		^{3}1200	(2400)	△	120
tetrahedron			(3600)	‖	600
3-lines					1200
Truncated Hyperdodecahedral					{10,6 |3 2}
decaprism	14400	28800	1440	⬭	720
tr icosidodecahedron			2400	⬭	120
tr octahedron			7200	□	600
hexaprism			(2400)	⬡	1200
nt^2t^2{10,6 |3 2}			(3600)	□	
Truncated Hypericosidodecatetraprism					
5-lines	1440	3600	2400	⬭	720
icosahedron		^{5}720	(2400)	△	120
tr tetrahedron			(3600)	‖	600
hexagon					1200
Truncated Icosahypertetrahedral					{6,10 |3 2}

Table 4: Icosapentaskews

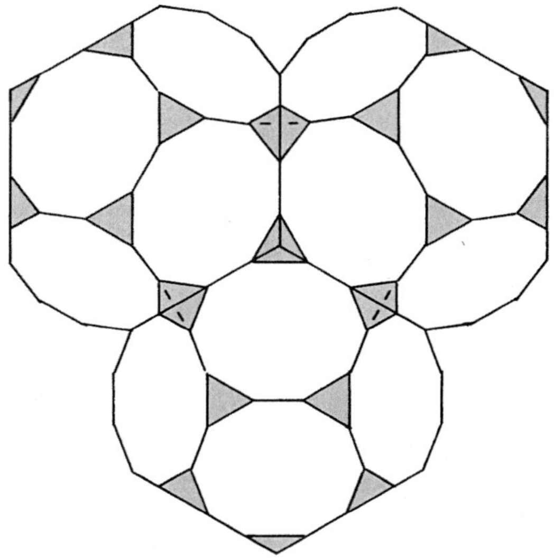

{10,6 |3 2} [t{5,3,3}]
Truncated
Hyperdodecahedral
Icosapentaskew

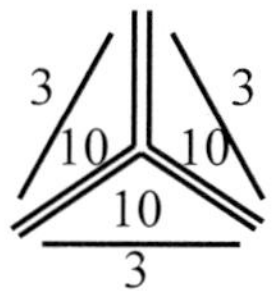

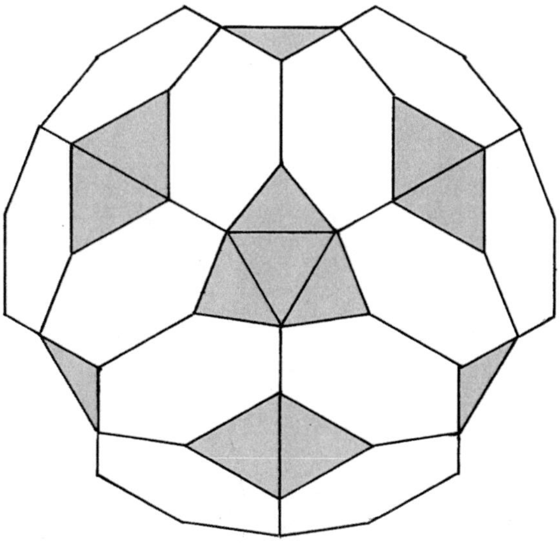

{6,10 |3 2} [t{3,3,5}]
Truncated
Icosahypertetrahedral
Icosapentaskew

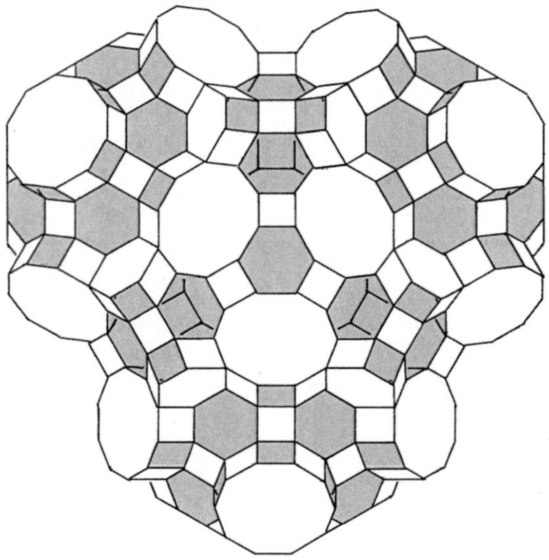

nt^2t^2{10,6 |3 2} [o{5,3,3}]
Truncated
Hypericosidodecatetraprism
Icosapentaskew

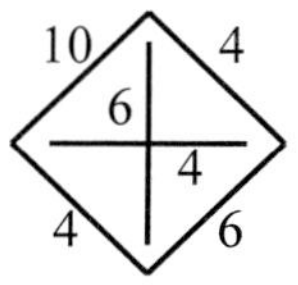

Fig.9: Icosapentaskews

MORE ICOSAPENTASKEWS

Solids	V	E	F	S
decagon	7200	10800	1440 ○	720
tr icosidodecahedron		^{2}3600	2400 △	120
tr tetrahedron			3600 □	600
triprism		(3600) ‖	(2400) ⬡	1200
Truncated Hypericosidodecatetrahedral				
pentagon	3600	7200	1440 ⬠	720
rhombicosidodecahedron			2400 △	120
octahedron		^{2}3600	3600 □	600
triprism		(3600) ‖	(2400) △	1200
Hyperhombicosidodecatetrahedral				
pentaprism	7200	18000	1440 ⬠	720
rhombicosidodecahedron			2400 ⬡	120
tr tetrahedron			3600 □	600
hexaprism			(2400) △	1200
			(3600) □	
Hyperhombicosidodecatetraprism				

Table 5: More Icosapentaskews

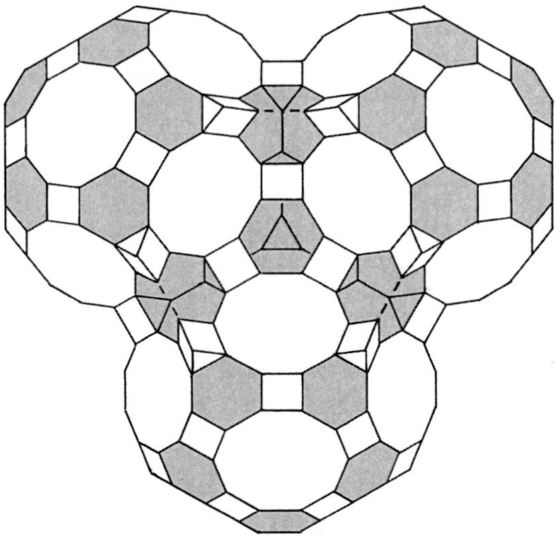

$[tt^2\{5,3,3\}]$

Truncated Hypericosi-
dodecatetrahedral
Icosapentaskew

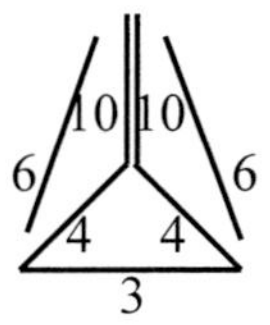

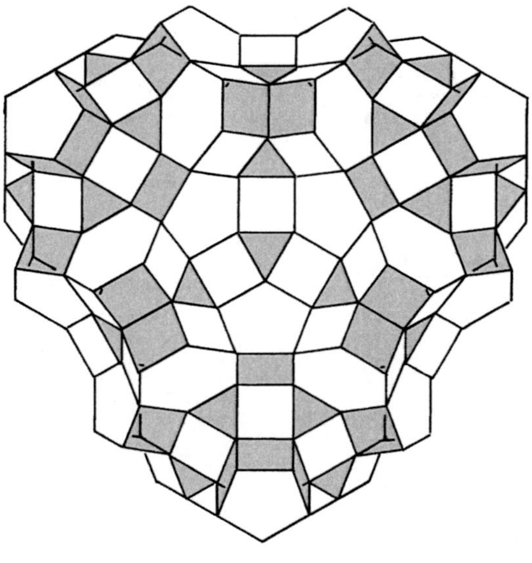

$[o^2\{5,3,3\}]$

Hyperhombicosi-
dodecatetraprism
Icosapentaskew

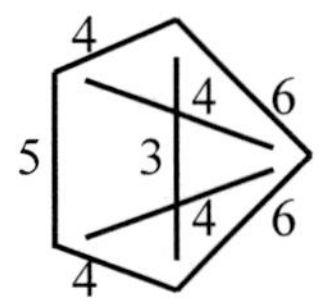

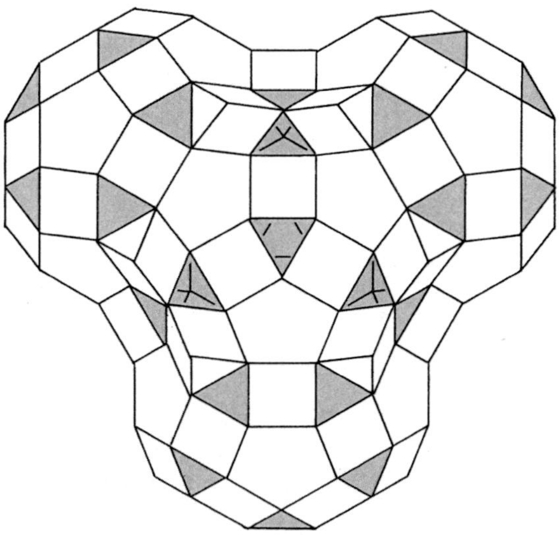

$[t^2t^2\{5,3,3\}]$

Hyperhombicosi-
dodecatetrahedral
Icosapentaskew

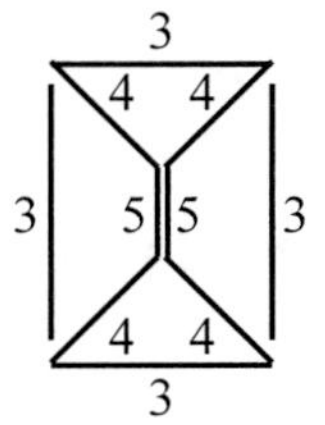

Fig.10: More Icosapentaskews

EVEN MORE ICOSAPENTASKEWS

Solids	V	E	F	S
pentaprism	7200	10800	1440 ⬠	720
tr icosahedron		^{2}3600	2400 ⬡	120
tr octahedron			3600 □	600
hexagon		(3600) ‖	(2400) ⬡	1200
Truncated Hypericositetratetrahedral				
pentaprism	3600	7200	1440 ⬠	720
icosidodecahedron		^{2}3600	2400 △	120
cuboctahedron			3600 □	600
triangle		(3600) ‖	(2400) △	1200
Hyperhombicositetratetrahedral				
decaprism	7200	18000	1440 ◯	720
tr dodecahedron			2400 △	120
cuboctahedron			3600 □	600
triprism			(2400) △	1200
			(3600) □	
Hyperhombicositetratetraprism				

Table 6: Even More Icosapentaskews

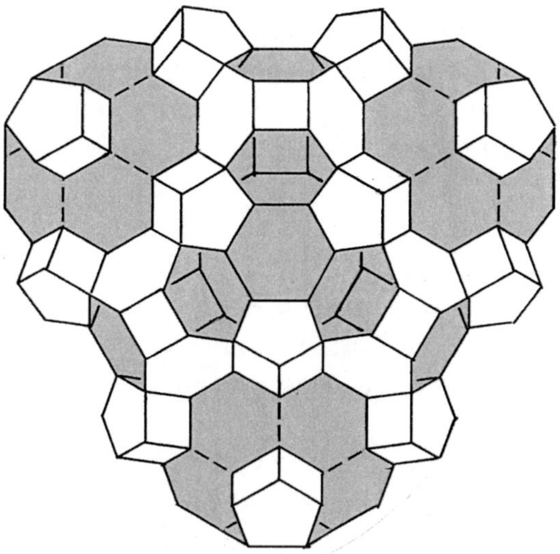

[tt^2{3,3,5}]
Truncated Hypericosi-
tetratetrahedral
Icosapentaskew

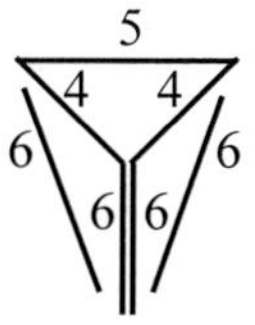

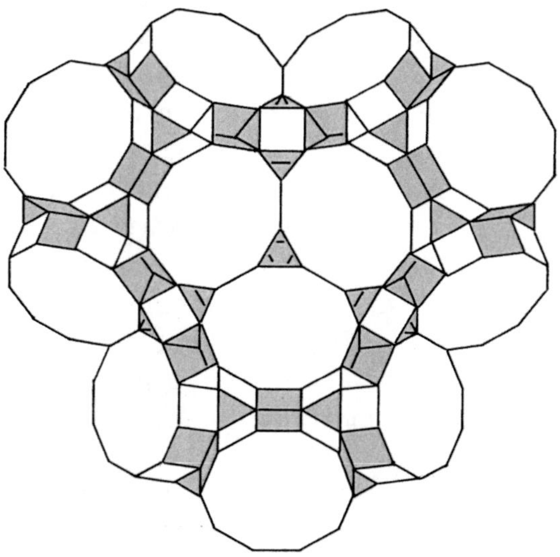

[o^2{3,3,5}]
Hyperhombicosi-
tetratetraprism
Icosapentaskew

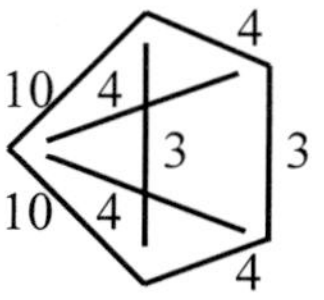

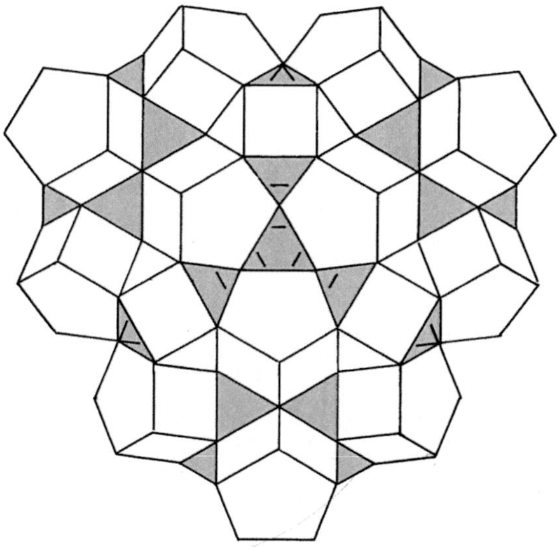

[t^{2}t^2{3,3,5}]
Hyperhombicosi-
tetratetrahedral
Icosapentaskew

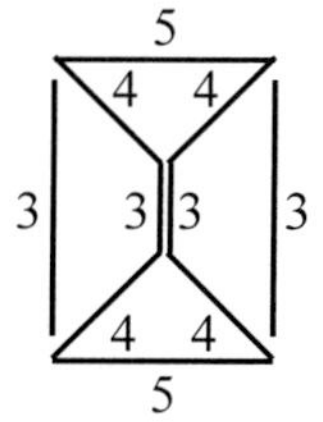

Fig.11: Even More Icosapentaskews

TRIACONTAHEXASKEW GRID

Finally taking the third possible alignment of the central tetrahedral vertex figure of o{5,3,3}, we can miss out the square faces of the truncated icosidodecahedra, making them 30-connected.

These link to 3-connected hexagon prisms to form one space, whilst the truncated octahedra, with missing square faces, are 6-connected to form the other space linked by 5-connected decagon prisms.

We have accordingly called these ‘Triacontahexaskews’ and again this vertex can be transformed to produce another full grid of nine members as shown adjoining in Figure 12.

All of these are again topologically similar surfaces between a 30-connected space and a 6-connected one. Taking double vertices and edges into account, they also have a consistent Euler Number (V-E+F) of -4560, interestingly the sum of the Euler Numbers of the previous two families.

Two of the resultants in this grid are also based on t^3 {5,3,3} = t^3 {3,3,5} and o^3 {5,3,3} = o^3 {3,3,5} respectively, and along with the central member are shown opposite in Figure 13, with Table 7 describing their numerical quantities.

Figure 14 and Table 8 on page 18 show the top left members of this grid, two intermediates based on the o^2 {3,3,5} and tt^2 {5,3,3}, and the resultant in the corner based on t{5,3,3}.

Similarly Figure 15 and Table 9 on page 19 illustrate the bottom right members of this grid, with its two intermediate forms based on tt^2 {3,3,5} and o^2 {5,3,3}, and the resultant in the corner based on t{3,3,5}.

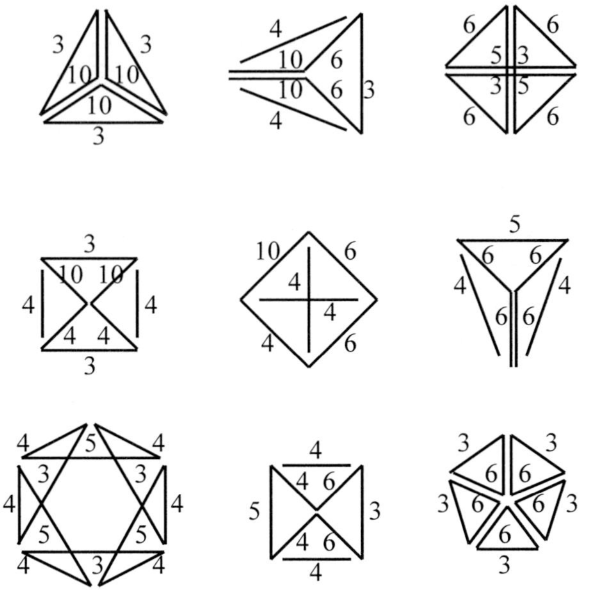

Fig.12: Triacontahexaskew Grid [5-6,3-30]

INCOMPLETE GRIDS

We have thus far examined three grids of vertex figures covering dodecatetraskews, icosapentaskews and triacontahexaskews. Only the first of these has intermediate or resultant members with tetrahedral vertex figures, which might produce further grids. On closer examination of these vertex figures we find that the two intermediates produce only the row or column in which they sit because the odd valued faces on their diagonals, {3} and {5}, cannot be produced by truncation.

The top row thus forms an incomplete grid with connectivity [4-4,2-12], whilst the right hand column forms another with connectivity [2-4,12-12]. The top right parent member of this dodecatetraskew grid has both diagonals with odd values, again {3} and {5}, which cannot be derived by truncation. It thus remains alone as an example of [4-4,12-12] connectivity.

TRIACONTAHEXASKEWS

Solids	V	E	F		S
tr tetrahedron	3600	²7200	1440	⬠	600
tr icosahedron			2400	⬡	120
pentagon			2400	△	720
triangle			(7200)	‖	1200
Hypertruncated Icositetrahedral					
tr octahedron	14400	28800	1440	○	600
tr icosidodecahedron			2400	⬡	120
decaprism			2400	⬡	720
hexaprism			3600	□	1200
			(7200)	□	
Truncated Hypericosidodecatetraprism					
tetrahedron	2400	²7200	1440	⬠	600
dodecahedron			2400	△	120
pentaprism			3600	□	720
triprism			(7200)	‖	1200
Hyperdodecatetraprism					

Table 7: Triacontahexaskews

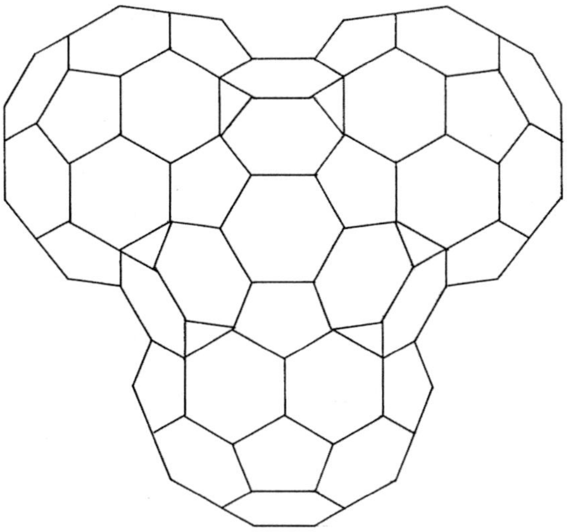

$[t^3\{5,3,3\}]$

Hypertruncated
Icositetrahedral
Triacontahexaskew

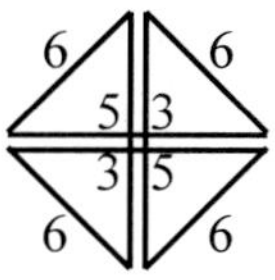

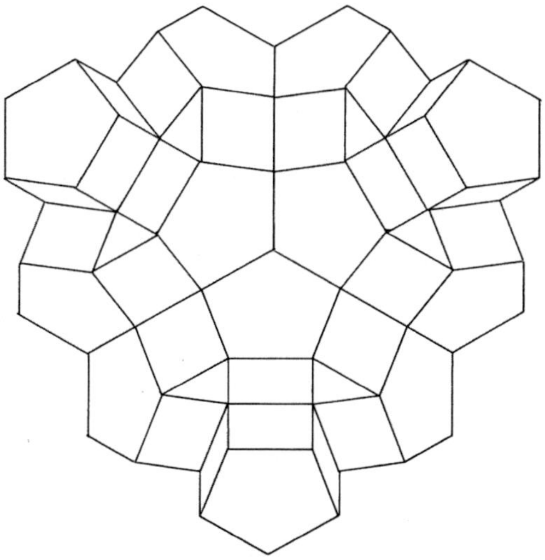

$[o^3\{5,3,3\}]$

Hyperdodecatetraprism
Triacontahexaskew

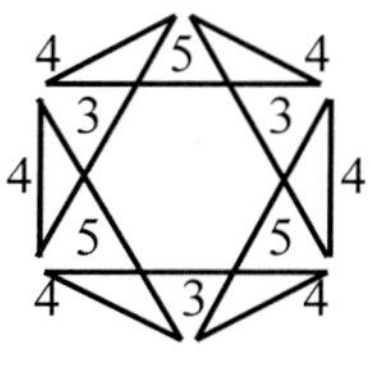

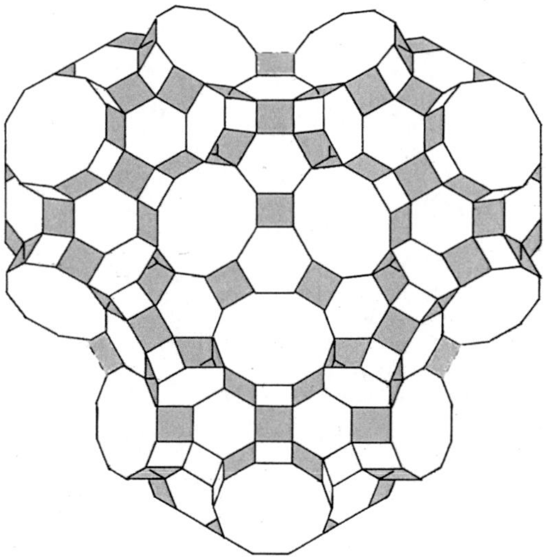

$[o\{5,3,3\}]$

Truncated Hypericosi-
dodecatetraprism
Triacontahexaskew

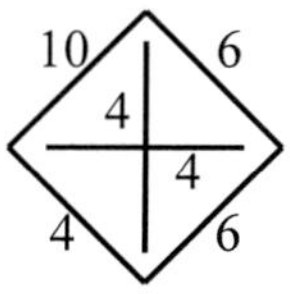

Fig.13: Triacontahexaskews

MORE TRIACONTAHEXASKEWS

Solids	V	E	F	S
tr tetrahedron	7200	10800	1440 ◯	600
tr icosidodecahedron		^{2}3600	2400 ⬡	120
decagon			2400 △	720
triprism		(3600) ∥	(3600) □	1200
Truncated Hypericosidodecatetrahedral				
tetrahedron	2400	^{2}3600	1440 ◯	600
tr dodecahedron		^{3}1200	2400 △	120
decagon			(3600) ∥	720
3-lines			(3600) ∥	1200
Truncated Hyperdodecahedral				
cuboctahedron	7200	14400	1440 ◯	600
tr dodecahedron		^{2}3600	2400 △	120
decaprism			2400 △	720
triprism			3600 □	1200
		(3600) ∥	(3600) □	
Hyperhombicositetratetraprism				

Table 8: More Triacontahexaskews

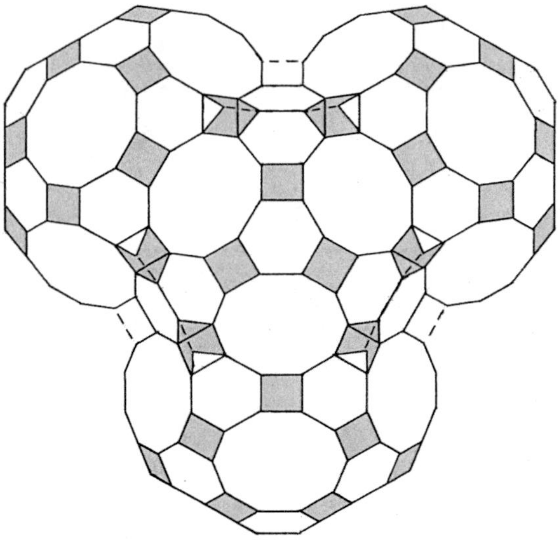

$[tt^2\{5,3,3\}]$
Truncated Hypericosi-
dodecatetrahedral
Triacontahexaskew

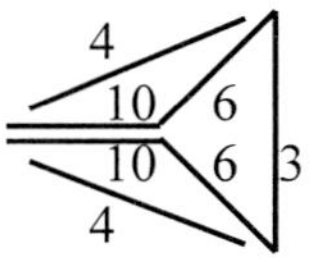

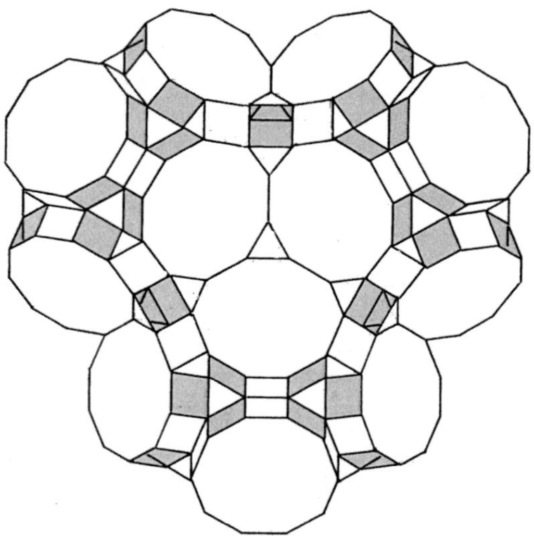

$[o^2\{3,3,5\}]$
Hyperhombicosi-
tetratetraprism
Triacontahexaskew

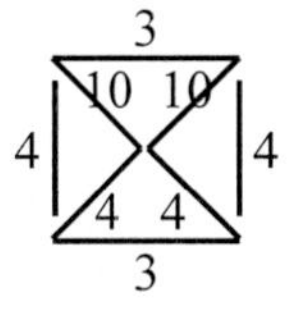

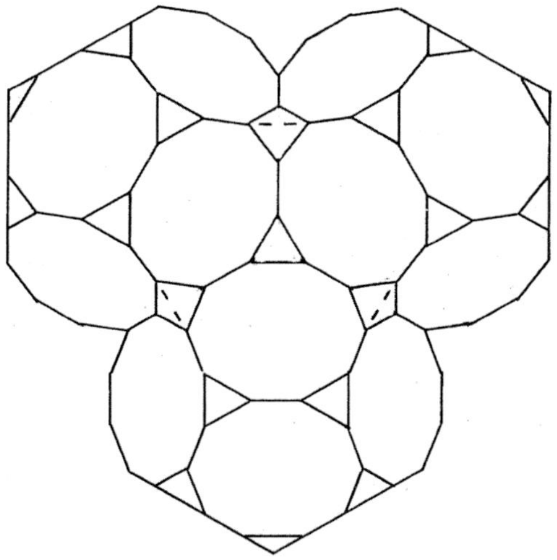

$[t\{5,3,3\}]$
Truncated
Hyperdodecahedral
Triacontahexaskew

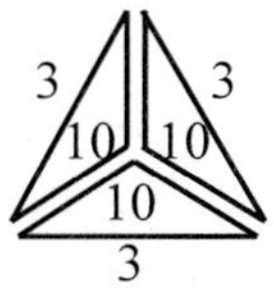

Fig.14: More Triacontahexaskews

EVEN MORE TRIACONTAHEXASKEWS

Solids	V	E	F	S
tr octahedron	7200	10800	1440 ⬠	600
tr icosahedron		^{2}3600	2400 ⬡	120
pentaprism			2400 ⬡	720
hexagon		(3600) ‖	(3600) □	1200
Truncated Hypericositetratetrahedral				
tr tetrahedron	1440	^{2}3600	2400 ⬡	600
icosahedron		^{5}720	2400 △	120
5-lines			(3600) ‖	720
hexagon			(3600) ‖	1200
Truncated Icosahypertetrahedral				
tr tetrahedron	7200	14400	1440 ⬠	600
rhombicosidodecahedron			2400 ⬡	120
pentaprism		^{2}3600	2400 △	720
hexaprism			3600 □	1200
		(3600) ‖	(3600) □	
Hyperhombicosidodecatetraprism				

Table 9: Even More Triacontahexaskews

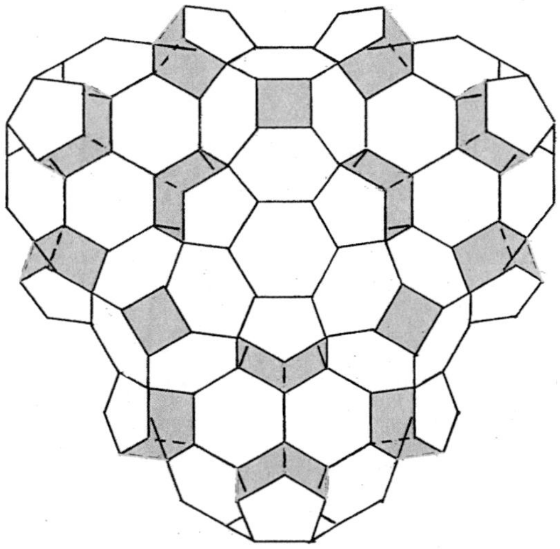

[tt^2{3,3,5}]
Truncated Hypericosi-
tetratetrahedral
Triacontahexaskew

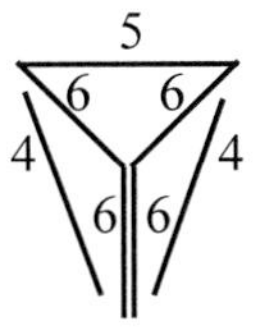

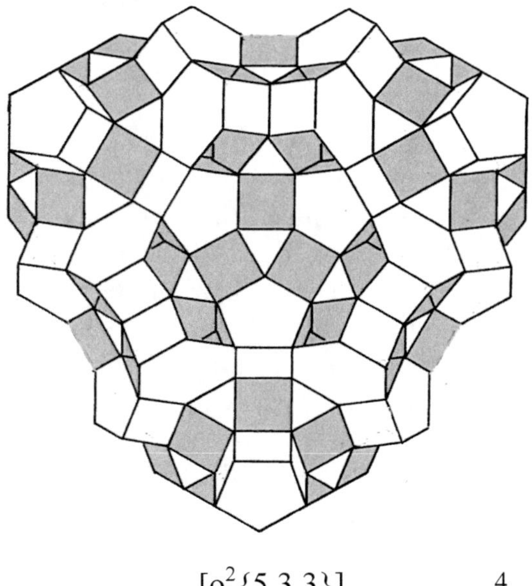

[o^2{5,3,3}]
Hyperhombicosi-
dodecatetraprism
Triacontahexaskew

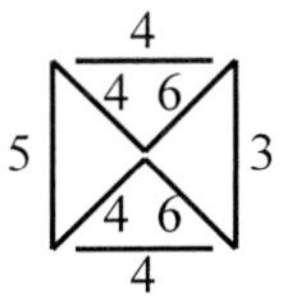

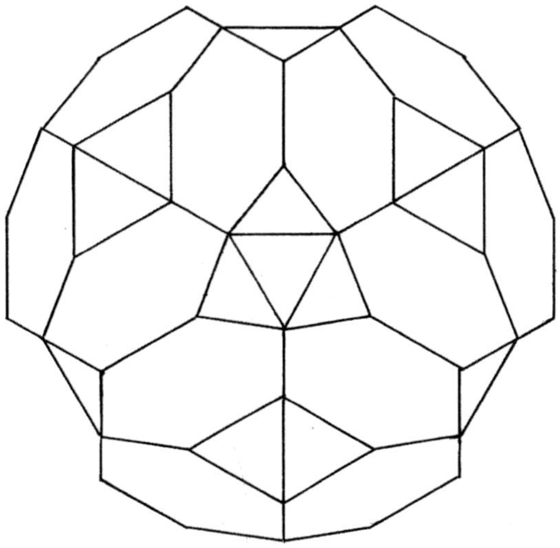

[t{3,3,5}]
Truncated
Icosahypertetrahedral
Triacontahexaskew

Fig.15: Even More Triacontahexaskews

THREE MORE FAMILY GRIDS

Now that we have examined the relatively complex families of hyperskews based on the Hyperdodecahedron {5,3,3} and Icosahypertetrahedron {3,3,5} family of hypersolids, we next come to look at those based on the simpler Hypercube {4,3,3} and Octahypertetrahedron {3,3,4} family.

Although the numbers are now more comprehensible and the hypercube itself perhaps capable of mental imagining, because these families are so much more folded up into four dimensions, distortions creep in all the sooner.

Views of a large cluster of solids as shown for the Dodecatetraskews etc. will not be so easily drawn, so the illustrations here are based on two pairs of cells from each space, effectively forming a tetrahedron of cells.

As before, the parent hyperskews {6,4 |4 3} and {4,6 |4 3}, based on the t^3 and o^3{4,3,3} respectively, have a common 'even faced' stretched version based on the tetrahedral vertex figure of o{4,3,3}. This last has truncated cuboctahedra with the octagon faces missing, connecting each six ways to 2-connected octagon prisms.

The other space here comprises truncated octahedra with half their hexagon faces missing, connecting four ways to 2-connected hexagon prisms. The two parents and their shared stretched version are shown opposite in Figure 17 and described numerically in Table 10.

Figure 18 and Table 11 on page 22 show the bottom right members of this grid, two intermediates based on the tt^2 and o^2{4,3,3} and the resultant in the corner based on t^2t^2{4,3,3}.

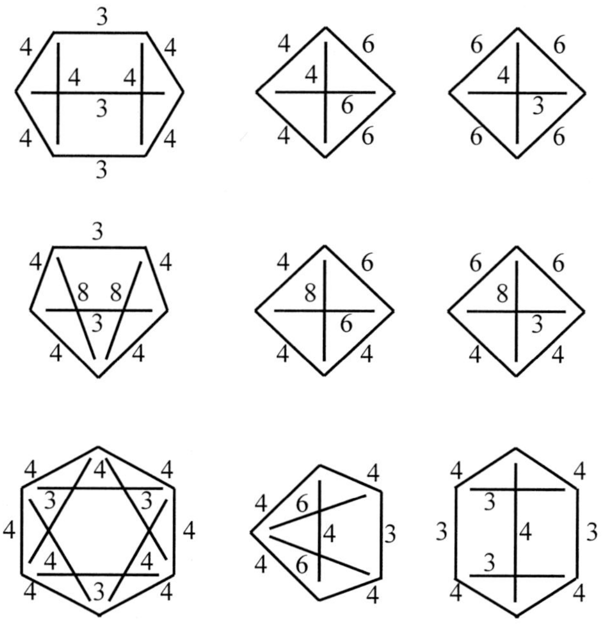

Fig.16: Hexatetraskew Grid [2-4,2-6]

Similarly Figure 19 and Table 12 on page 23 show the top left members of this grid based on tt^2, o^2 and t^2t^2{3,3,4}.

The smaller numbers tabulated in this family emphasise the fact that all the hyperskews are indeed finite assemblies of polygons that actually close up in four dimensions. This particular family are again all topologically similar surfaces separating two different linked spaces, with this time a Euler Number (V-E+F) of -32.

Figure 16 above shows the full grid for this family, all members of which comprise the same two spaces, one 6-connected the other 4-connected, hence 'Hexatetraskews'. Again these are all produced by the process of vertex transformation as was described in *'The Skew Polyhedra'*.

HEXATETRASKEWS

Solids	V	E	F	S
tr tetrahedron	96	192	64 ⬡	16
tr octahedron			(32) △	8
{6,4 |4 3}			(24) □	
Hypertruncated Octetrahedral				
tr octahedron	384	768	64 ⬡	16
tr cuboctahedron			288 □	8
octaprism			(64) ⬡	24
hexaprism			(48) ⯃	32
nt^2t^2 {4,6 |4 3}				
Truncated Hypercuboctetraprism				
tetrahedron	64	192	96 □	16
cube			(64) △	8
cube			(48) □	24
triprism				32
{4,6 |4 3}				
Hyperhexatetraprism				

Table 10: Hexatetraskews

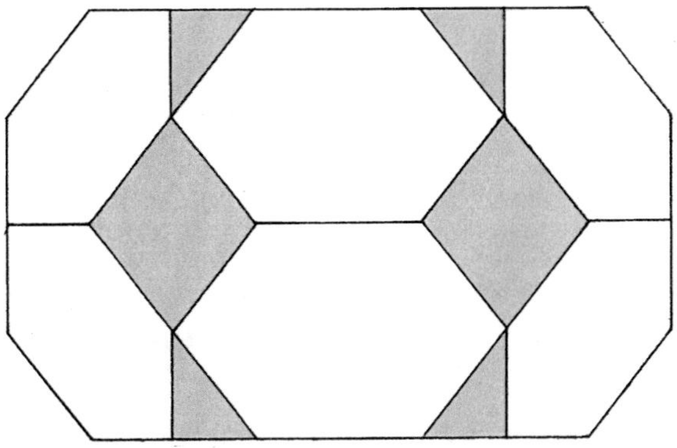

{6,4 |4 3} [t^3{4,3,3}]
Hypertruncated
Octetrahedral
Hexatetraskew

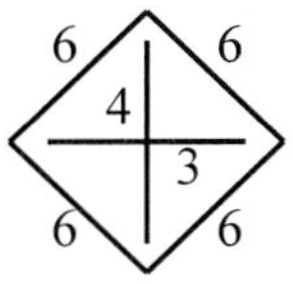

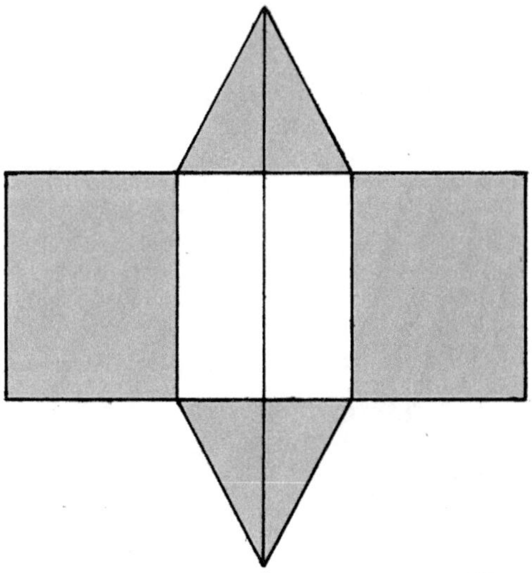

{4,6 |4 3} [o^3{4,3,3}]

Hyperhexatetraprism
Hexatetraskew

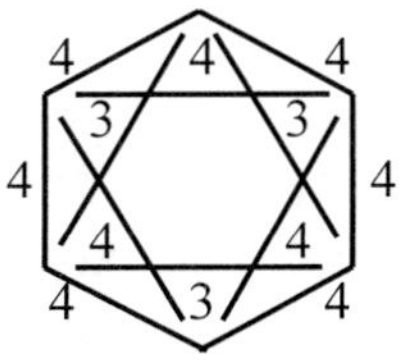

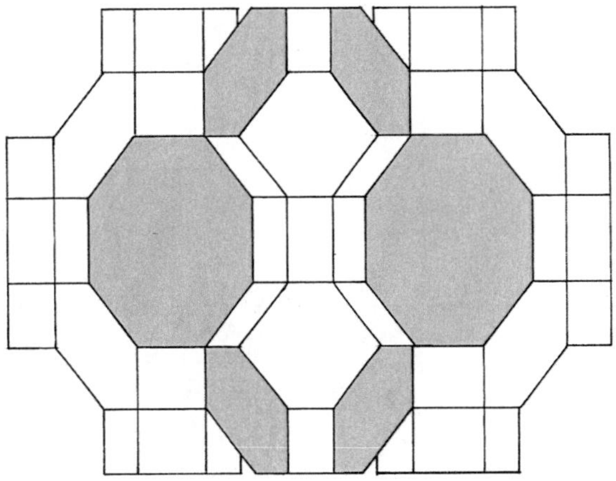

nt^2t^2{4,6 |4 3} [o{4,3,3}]
Truncated
Hypercuboctetraprism
Hexatetraskew

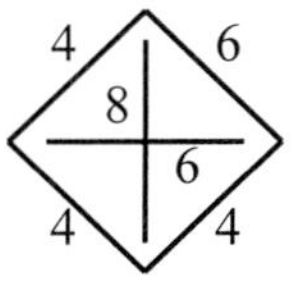

Fig.17: Hexatetraskews

MORE HEXATETRASKEWS

Solids	V	E	F	S
tr tetrahedron	192	384	64 ⬡	16
tr cuboctahedron			96 □	8
triprism			(64) △	32
			(24) ⯃	
Truncated Hypercuboctetrahedral				
octahedron	96	288	64 △	16
rhombicuboctahedron			96 □	8
triprism			(64) △	32
			(24) □	
Hyperhombicuboctetrahedral				
tr tetrahedron	192	480	64 △	16
rhombicuboctahedron			192 □	8
cube			(64) ⬡	24
hexaprism			(48) □	32
Hyperhombicuboctetraprism				

Table 11: More Hexatetraskews

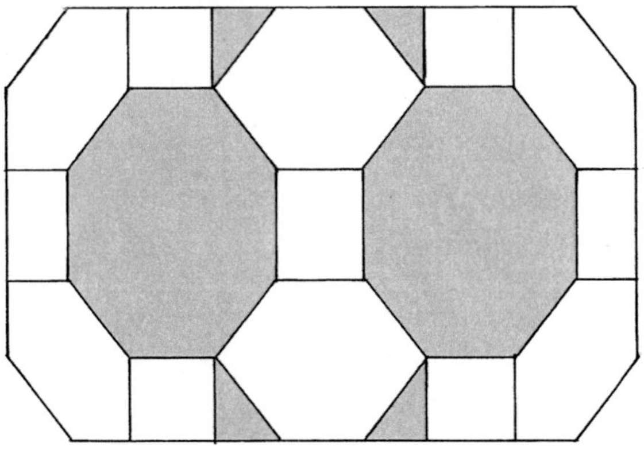

[tt²{4,3,3}]

Truncated Hyper-
cuboctetrahedral
Hexatetraskew

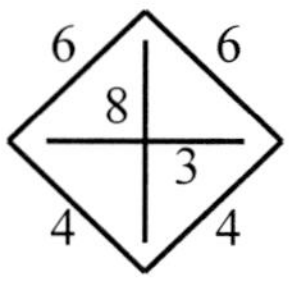

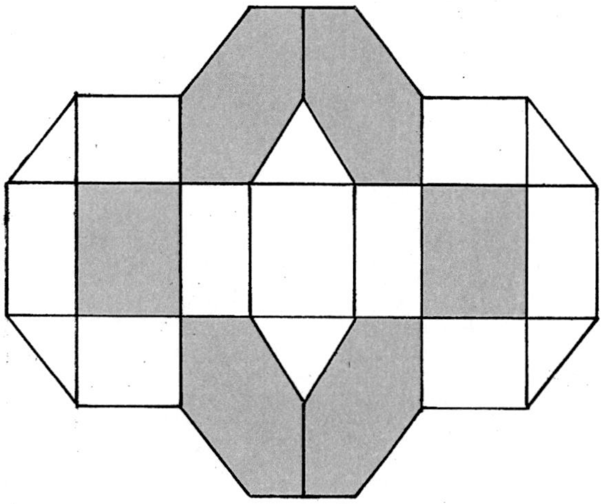

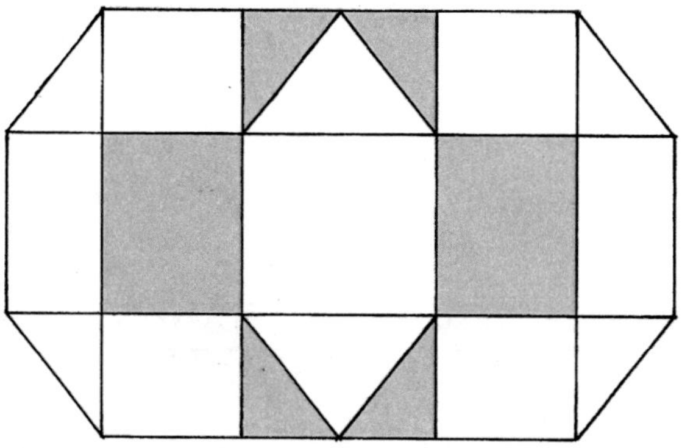

[o²{4,3,3}]

Hyperhombi-
cuboctetraprism
Hexatetraskew

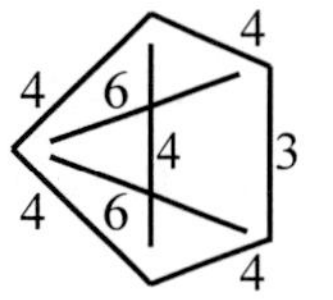

[t²t²{4,3,3}]

Hyperhombi-
cuboctetrahedral
Hexatetraskew

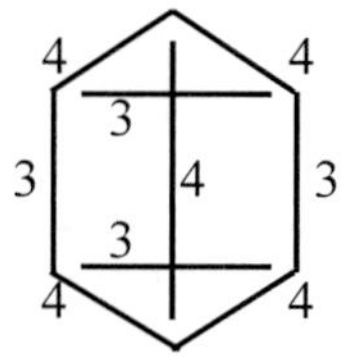

Fig.18: More Hexatetraskews

EVEN MORE HEXATETRASKEWS

Solids	V	E	F	S
tr octahedron	192	384	64 ⬡	16
tr octahedron			96 □	8
cube			(32) ⬡	24
			(48) □	
Truncated Hyperoctahedral				
cuboctahedron	96	288	64 △	16
cuboctahedron			96 □	8
cube			(32) △	24
			(48) □	
Hypercuboctahedral				
cuboctahedron	192	480	64 △	16
tr cube			192 □	8
octaprism			(64) △	24
triprism			(48) octagon	32
Hypercuboctaprism				

Table 12: Even More Hexatetraskews

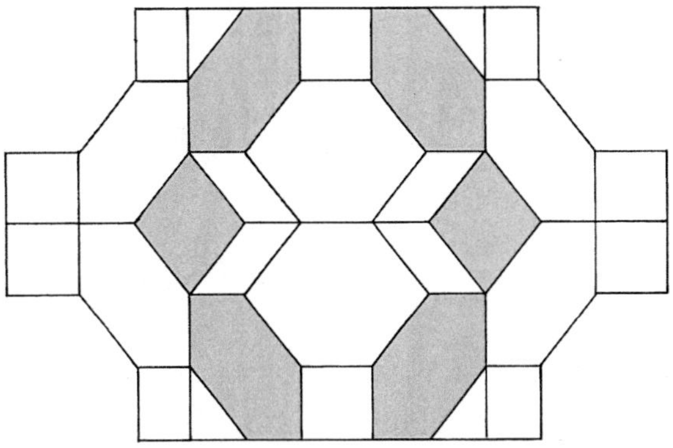

[tt^2{3,3,4}]

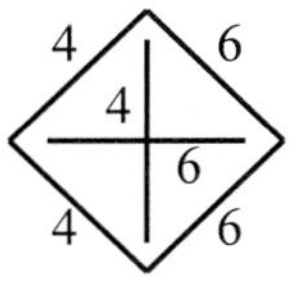

Truncated
Hyperoctahedral
Hexatetraskew

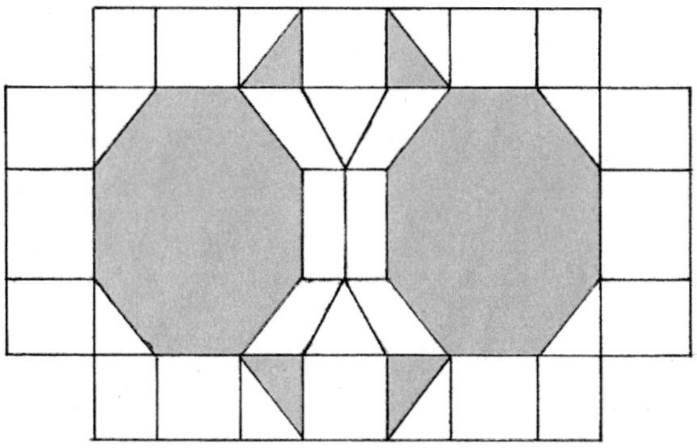

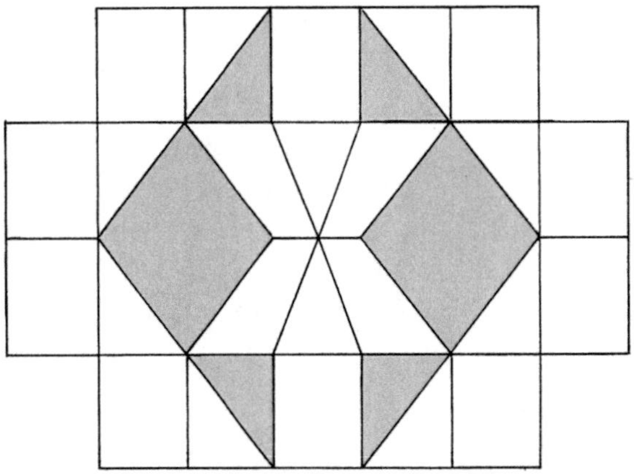

[o^2{3,3,4}]

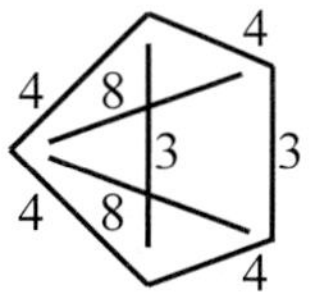

Hypercuboctaprism
Hexatetraskew

[t^{2}t^2{3,3,4}]

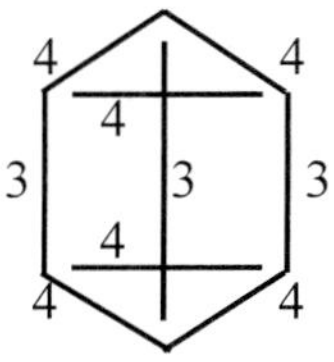

Hypercuboctahedral
Hexatetraskew

Fig.19: Even More Hexatetraskews

AN EXTRA FAMILY GRID

Like the dodecatetraskews, Figure 16's hexatetraskew grid has three further tetrahedral vertex figures that show some promise for producing further grids.

The right hand central intermediate has one odd missing face {3}, which cannot be produced by truncation and so will produce only the column it sits in as a further incomplete grid, this time with [2-4,6-6] connectivity.

The top central intermediate in Figure 16 also has a tetrahedral vertex figure, fortunately without any odd numbers involved, so the possibility of a further grid derived from this presents itself. This is shown adjoining in Figure 20, giving rise to the hexatetratetraskews, with [4-4,2-6] connectivity. These retain the same Euler Number of -32 as the hexatetraskews, but account has to be taken of double edges and multiple faces.

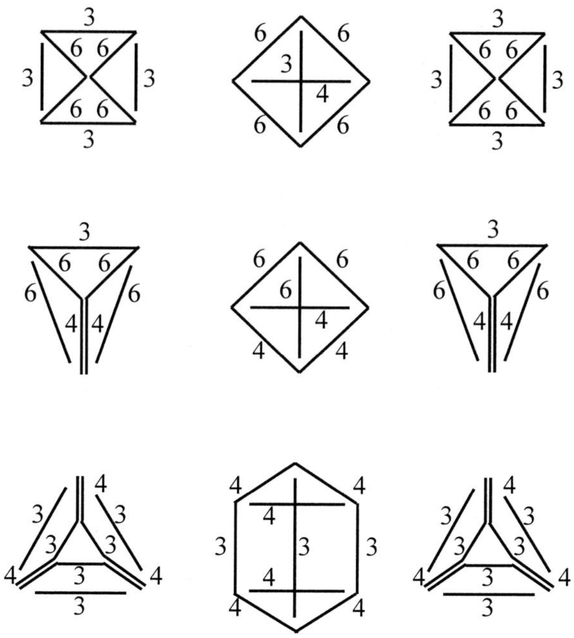

Fig.20: Hexatetratetraskew Grid [4-4,2-6]

Those in the central column of this grid already appeared as the top row of the hexatetraskew [2-4,2-6] grid in Figure 16, and have thus already been described. Furthermore, since the left hand column mirrors the right, there remain in fact just three new forms to describe. These are shown in Figure 21 opposite and their numerical quantities are set out in Table 13 adjoining.

The third outstanding tetrahedral vertex figure is the resultant at top right in Figure 16. This is also the central top member of the hexatetratetraskew [4-4,2-6] grid shown here in Figure 20. It has one odd missing face, and will yield another incomplete grid, which is in fact a repeat of the top row of Figure 20, so there is nothing new here other than its [4-4,6-6] connectivity.

HEXATETRATETRASKEWS

Solids	V	E	F	S
tr tetrahedron	48	96	32 ⬡	8
tr tetrahedron		^{2}24	32 △	8
octahedron			(32) △	8
			(24) ‖	
Truncated Octahypertetrahedral				
tr tetrahedron	96	144	32 ⬡	8
tr tetrahedron		^{2}48	32 △	8
tr octahedron			48 □	8
		(48) ‖	(32) ⬡	
Hypertruncated Octetrahedral				
tetrahedron	32	48	32 △	8
tetrahedron		^{2}48	48 □	8
cuboctahedron			(32) △	8
			(48) ‖	
Hypercuboctetrahedral				

Table 13: Hexatetratetraskews

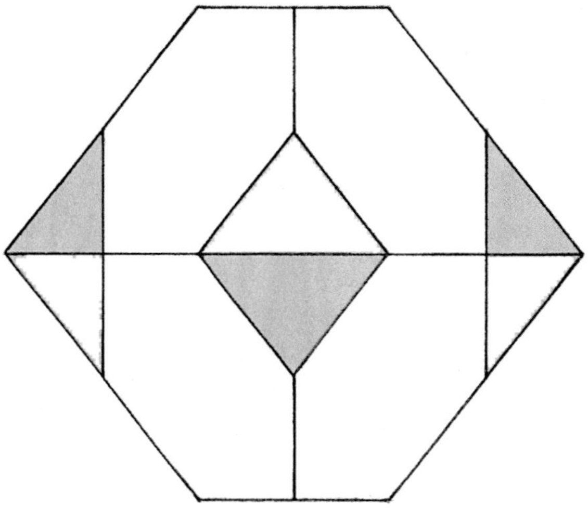

[t{3,3,4}]

Truncated Octahypertetrahedral Hexatetratetraskew

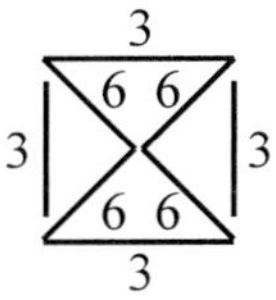

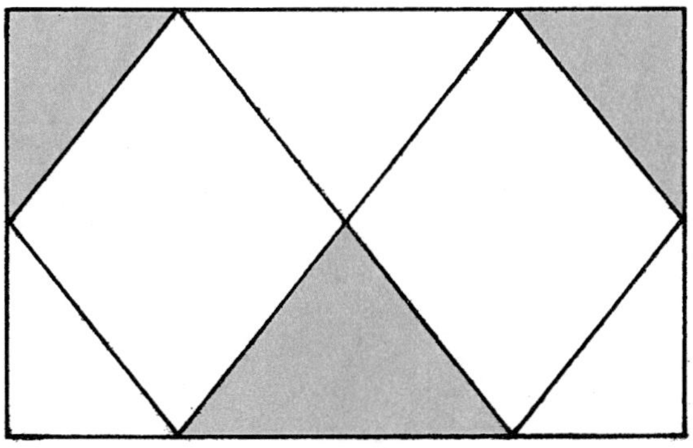

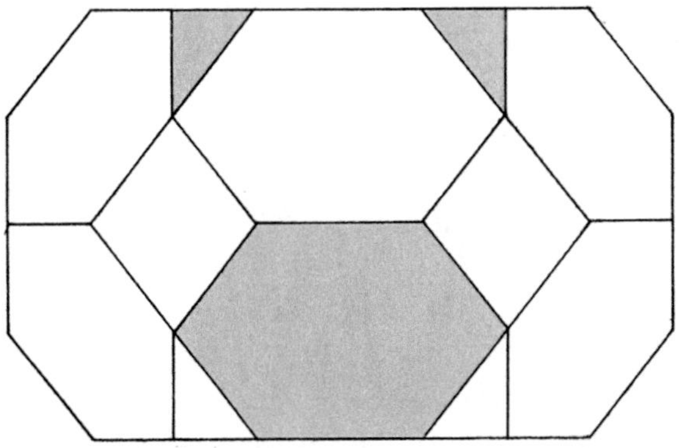

[t^{2}{4,3,3}]

Hypercuboctetrahedral Hexatetratetraskew

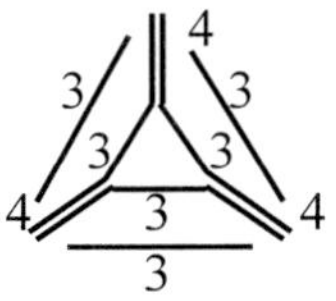

[t^{3}{3,3,4}]

Hypertruncated Octetrahedral Hexatetratetraskew

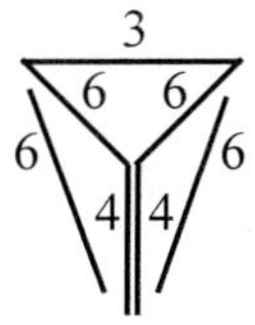

Fig.21: Hexatetratetraskews

OCTATETRASKEW GRID

Looking again at the central member of the hexatetraskew grid based on o{4,3,3}, we can as before realign this vertex in two further ways missing out different polygons and creating different connections.

The first realignment finds the family of ‘octatetraskews’, analogous to the icosapentaskews seen earlier. The grid for this family is shown adjoining in Figure 22. These will be found to have Euler Number (V-E+F) of -80, once account is taken of multiple faces and edges.

The central member has the truncated cuboctahedra with missing hexagons, 8-connected to 4-connected truncated octahedra. The other space here comprises 4-connected octaprisms and 3-connected hexaprisms.

As before the grid contains two notional ‘parent’ hyperskews which can be described with amended Coxeter notation {8,6 |3 2} and {6,8 |3 2}, top left and bottom right respectively. These are based on t{4,3,3} and t{3,3,4} respectively, and along with the central member are shown adjoining in Figure 23, with Table 14 describing their numerical quantities.

Figure 24 and Table 15 on page 28 show the bottom left members of this grid, two intermediates based on the tt^2 and o^2{4,3,3} and the resultant in the corner based on t^{2}t^2{4,3,3}.

Similarly Figure 25 and Table 16 on page 29 show the top right members of the same grid, two intermediates based on tt^2 and o^2 {3,3,4} and a resultant based on t^{2}t^2{3,3,4}.

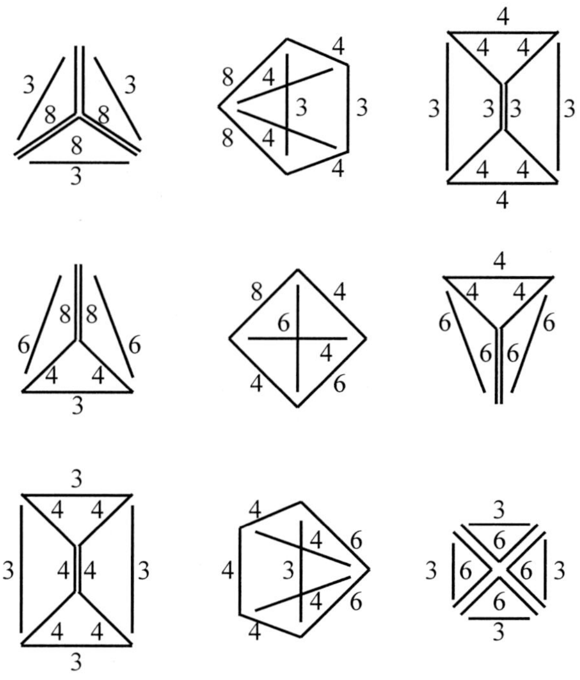

Fig.22: Octatetraskew Grid [3-4,4-8]

OCTATETRASKEWS

Solids	V	E	F	S
octagon	64	96	48 ⯃	24
truncated cube		$^{3}32$	(64) △	8
tetrahedron			(96) ∥	16
3-lines				32
Truncated Hypercubic			{8,6 |3 2}	
octaprism	384	768	48 ⯃	24
tr cuboctahedron			64 ⬡	8
tr octahedron			192 □	16
hexaprism			(64) ⬡	32
nt^2t^2{8,6 |3 2}			(96) □	
Truncated Hypercuboctetraprism				
4-lines	48	96	64 ⬡	24
octahedron		$^{4}24$	(64) △	8
tr tetrahedron			(96) ∥	16
hexagon				32
Truncated Octahypertetrahedral			{6,8 |3 2}	

Table 14: Octatetraskews

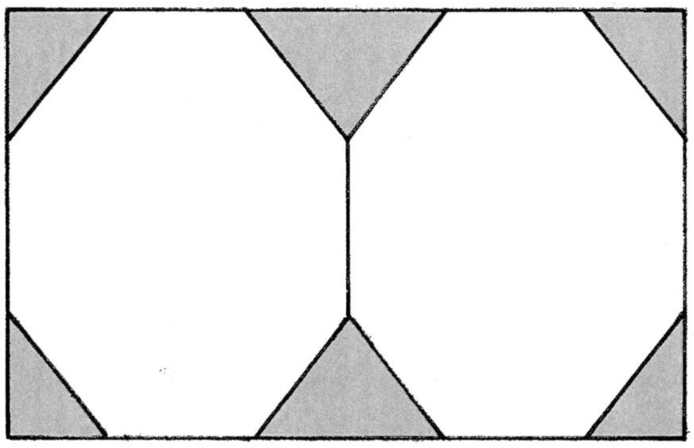

{8,6 |3 2} [t{4,3,3}]
Truncated
Hypercubic
Octatetraskew

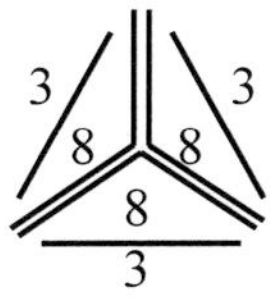

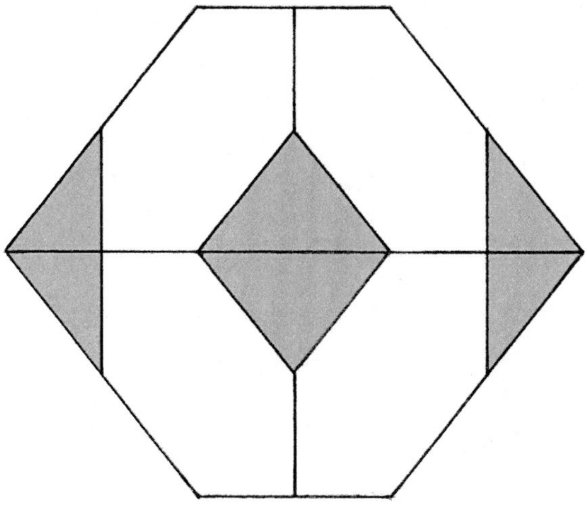

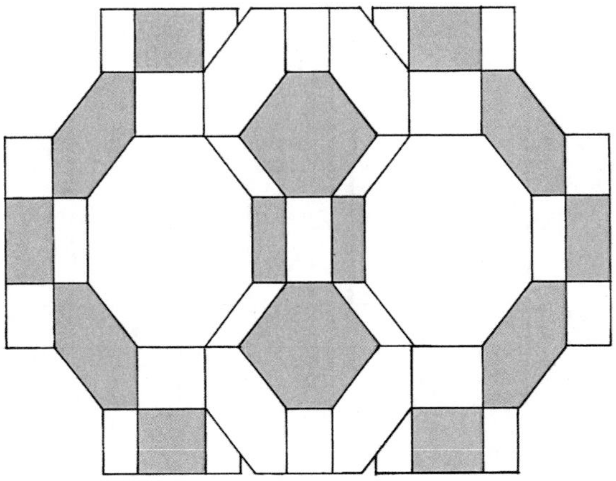

{6,8 |3 2} [t{3,3,4}]
Truncated
Octahypertetrahedral
Octatetraskew

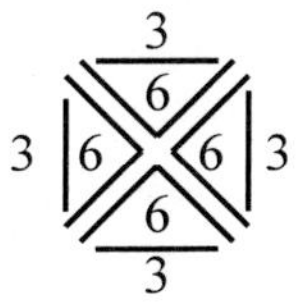

nt^2t^2{8,6 |3 2} [o{4,3,3}]
Truncated
Hypercuboctetraprism
Octatetraskew

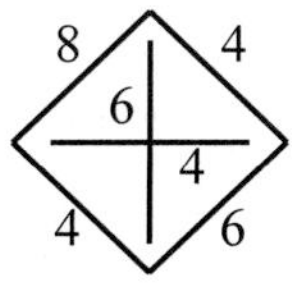

Fig.23: Octatetraskews

MORE OCTATETRASKEWS

Solids	V	E	F	S
octagon	192	288	48 (octagon)	24
tr cuboctahedron		^{2}96	64 △	8
tr tetrahedron			96 □	16
triprism		(96) ‖	(64) ⬡	32
Truncated Hypercuboctetrahedral				
square	96	192	48 □	24
rhombicuboctahedron		^{2}96	64 △	8
octahedron			96 □	16
triprism		(96) ‖	(64) △	32
Hyperhombicuboctetrahedral				
cube	192	480	48 □	24
rhombicuboctahedron			64 ⬡	8
tr tetrahedron			96 □	16
hexaprism			(64) △	32
			(96) □	
Hyperhombicuboctetraprism				

Table 15: More Octatetraskews

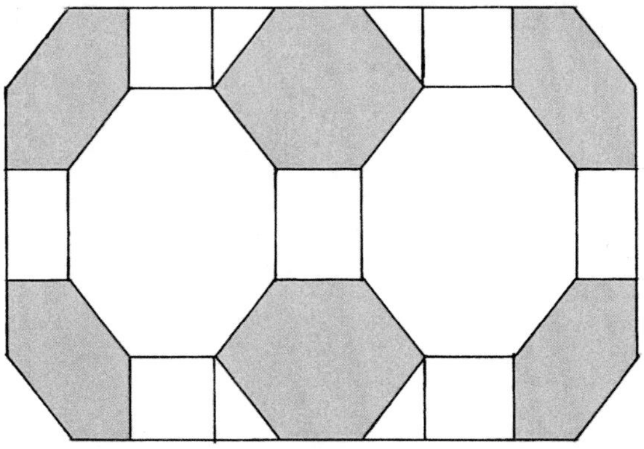

$[tt^2\{4,3,3\}]$

Truncated Hypercuboctetrahedral Octatetraskew

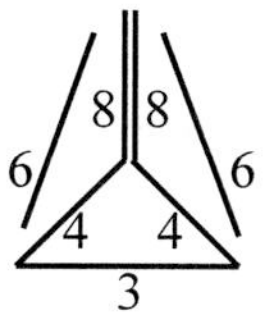

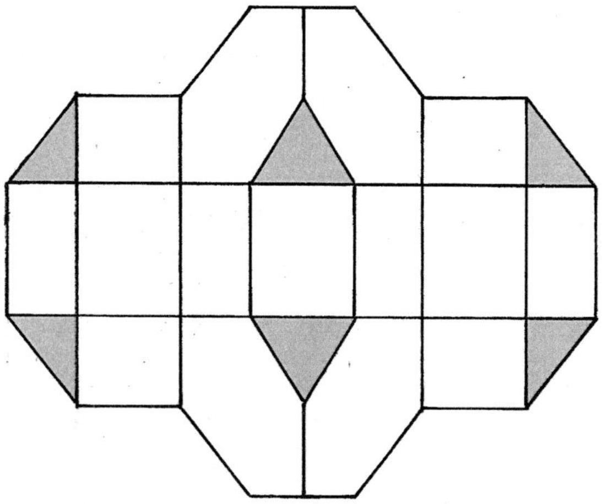

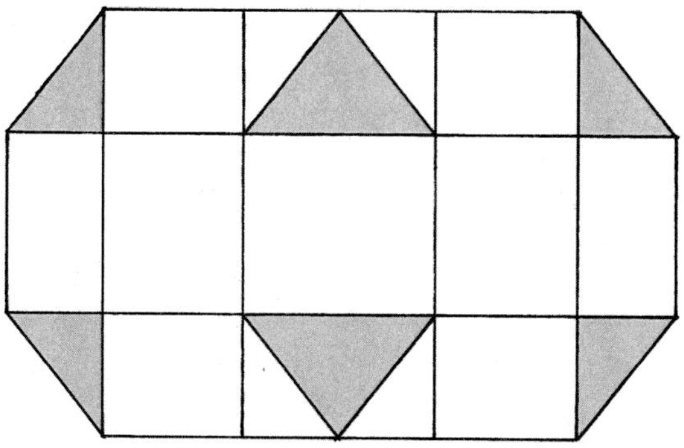

$[o^2\{4,3,3\}]$

Hyperhombi-cuboctetraprism Octatetraskew

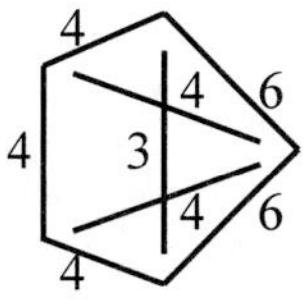

$[t^2t^2\{4,3,3\}]$

Hyperhombi-cuboctetrahedral Octatetraskew

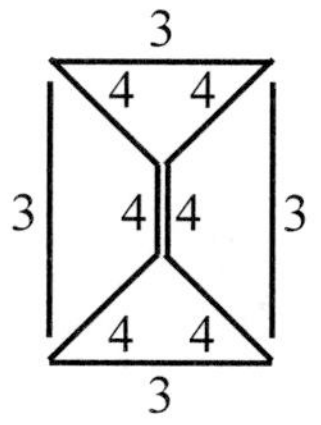

Fig.24: More Octatetraskews

EVEN MORE OCTATETRASKEWS

Solids	V	E	F	S
cube	192	288	48 □	24
tr octahedron		$^{2}96$	64 ⬡	8
tr octahedron			96 □	16
hexagon		(96) ∥	(64) ⬡	32
Truncated Hyperoctahedral				
cube	96	192	48 □	24
cuboctahedron		$^{2}96$	64 △	8
cuboctahedron			96 □	16
triangle		(96) ∥	(64) △	32
Hypercuboctahedral				
octaprism	192	480	48 ⯃	24
truncated cube			64 △	8
cuboctahedron			96 □	16
triprism			(64) △	32
			(96) □	
Hypercuboctaprism				

Table 16: Even More Octatetraskews

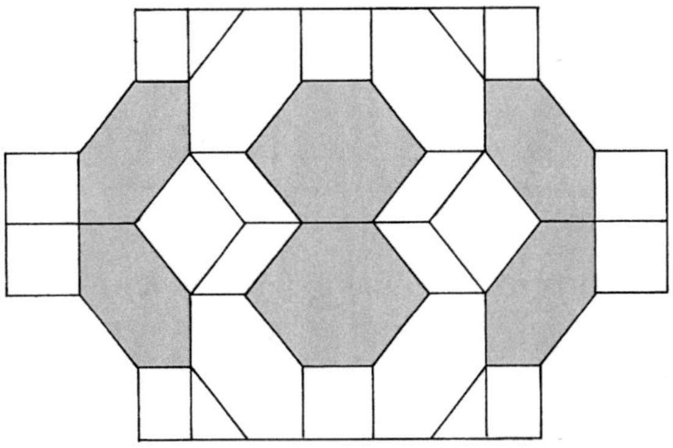

[tt²{3,3,4}]

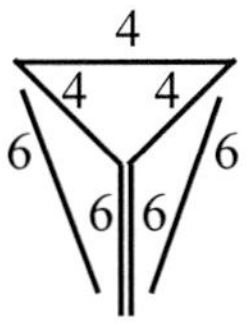

Truncated Hyperoctahedral Octatetraskew

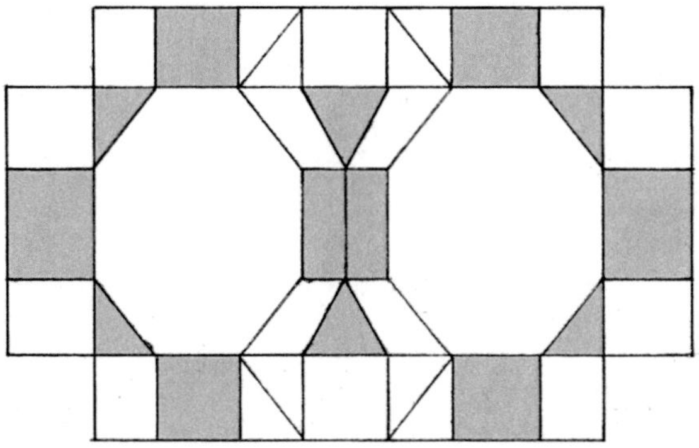

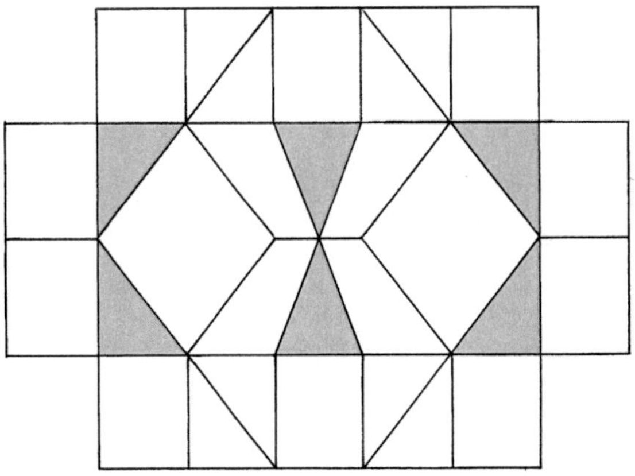

[o²{3,3,4}]

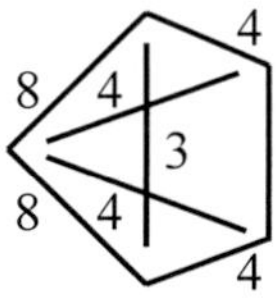

Hypercuboctaprism Octatetraskew

[t²t²{3,3,4}]

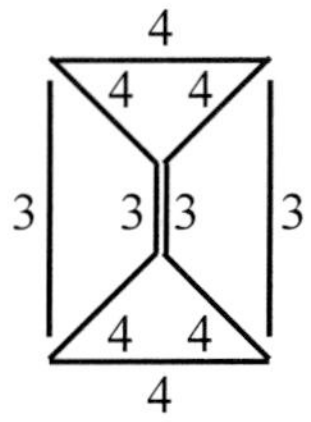

Hypercuboctahedral Octatetraskew

Fig.25: Even More Octatetraskews

DODECAHEXASKEW GRID

Returning again to our central member, its third and final alignment misses out two squares from the vertex figure, producing truncated cuboctahedra that are 12-connected to 3-connected hexaprisms. The other space created comprises truncated octahedra 6-connected to 4-connected octaprisms.

The other members of the grid produced by vertex transformation from this same central member share these same connections and are accordingly dubbed 'dodecahexaskews'. The full grid of nine members, analogous to the triacontahexaskews seen earlier, is shown adjoining in Figure 26.

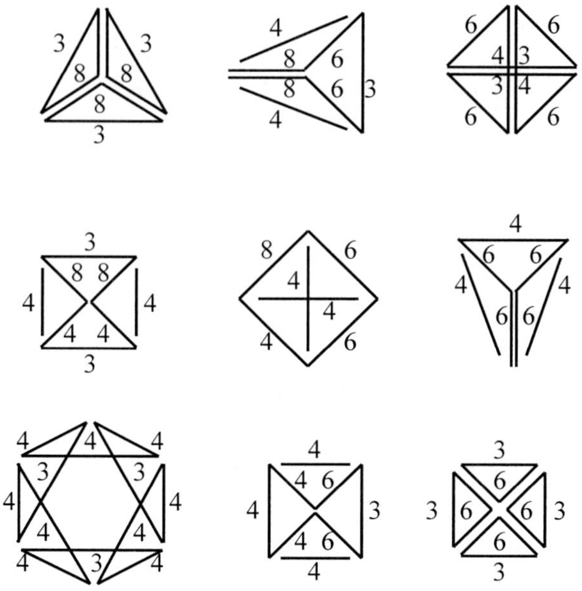

Fig.26: Dodecahexaskew Grid [4-6,3-12]

Since the members of this family are again topologically similar, taking account of multiple faces and edges they have a common Euler Number (V-E+F). This time it is -112, again the sum of those for the previous two families derived from the odd hypersolid o{4,3,3}.

Two of the resultants in this grid are also based on t^3{4,3,3} = t^3{3,3,4} and o^3{4,3,3} = o^3{3,3,4} respectively, and along with the central member are shown opposite in Figure 27, with Table 17 describing their numerical quantities.

Figure 28 and Table 18 on page 32 show the top left members of this grid, two intermediates based on the o^2{3,3,4} and tt^2 {4,3,3}, and the resultant in the corner based on t{4,3,3}.

Similarly Figure 29 and Table 19 on page 33 illustrate the bottom right members of this grid, with its two intermediate forms based on tt^2{3,3,4} and o^2{4,3,3}, and the resultant in the corner based on t{3,3,4}.

DODECAHEXASKEWS

Solids	V	E	F	S
tr tetrahedron	96	^{2}192	48 □	16
tr octahedron			64 △	8
square			64 ⬡	24
triangle			(192) ‖	32
Hypertruncated Octetrahedral				
tr octahedron	384	768	48 ⯃	16
tr cuboctahedron			64 ⬡	8
octaprism			64 ⬡	24
hexaprism			96 □	32
			(192) □	
Truncated Hypercuboctetraprism				
tetrahedron	64	^{2}192	48 □	16
cube			64 △	8
cube			96 □	24
triprism			(192) ‖	32
Hyperhexatetraprism				

Table 17: Dodecahexaskews

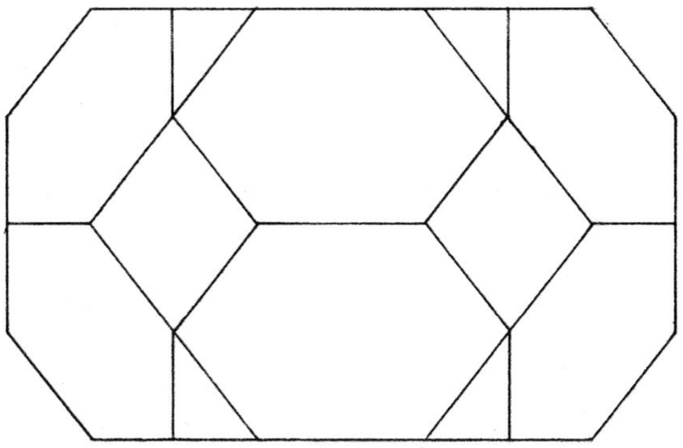

$[t^3\{4,3,3\}]$

Hypertruncated
Octetrahedral
Dodecahexaskew

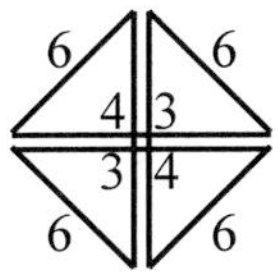

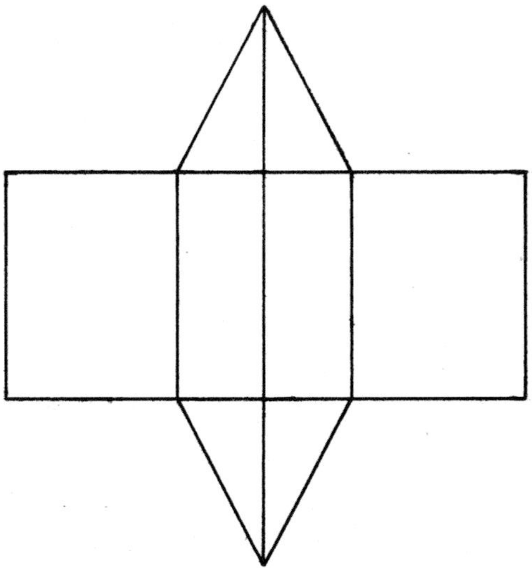

$[o^3\{4,3,3\}]$

Hyperhexatetraprism
Dodecahexaskew

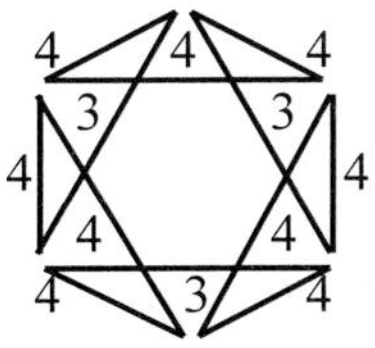

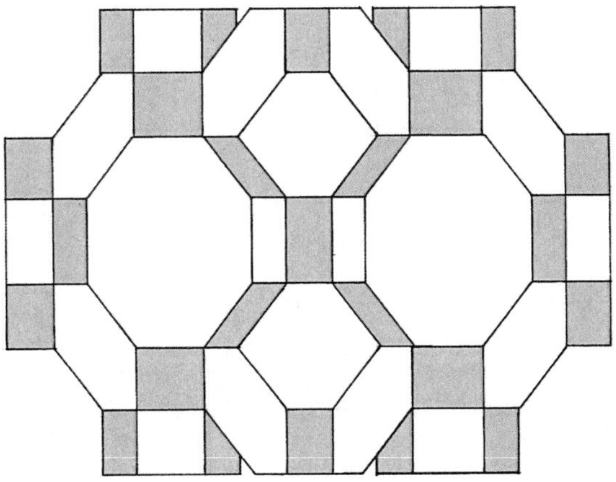

$[o\{4,3,3\}]$

Truncated
Hypercuboctetraprism
Dodecahexaskew

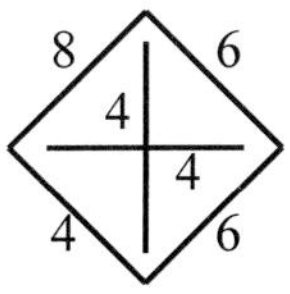

Fig.27: Dodecahexaskews

MORE DODECAHEXASKEWS

Solids	V	E	F	S
tr tetrahedron	192	288	48 ⯃	16
tr cuboctahedron		^{2}96	64 ⬡	8
octagon			64 △	24
triprism		(96) ‖	(96) □	32
Truncated Hypercuboctetrahedral				
tetrahedron	64	^{2}96	48 ⯃	16
truncated cube		^{3}32	64 △	8
octagon			(96) ‖	24
3-lines			(96) ‖	32
Truncated Hypercubic				
cuboctahedron	192	384	48 ⯃	16
truncated cube		^{2}96	64 △	8
octaprism			64 △	24
triprism			96 □	32
		(96) ‖	(96) □	
Hypercuboctaprism				

Table 18: More Dodecahexaskews

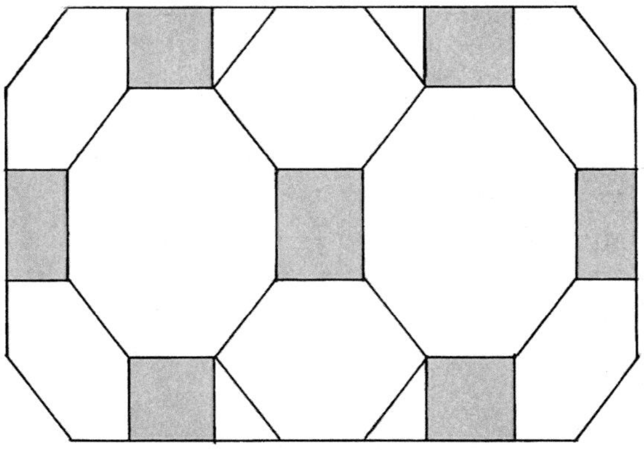

[tt^2{4,3,3}]

Truncated
Hypercuboctetrahedral
Dodecahexaskew

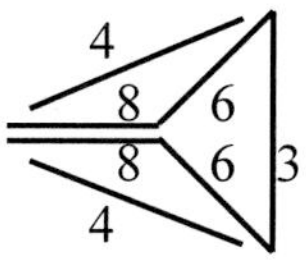

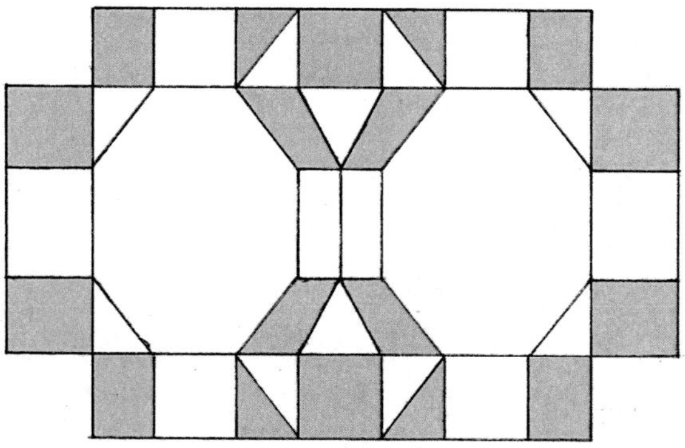

[o^2{3,3,4}]

Hypercuboctaprism
Dodecahexaskew

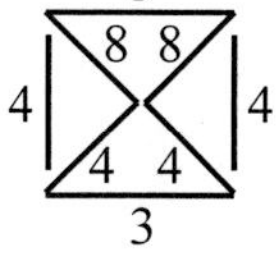

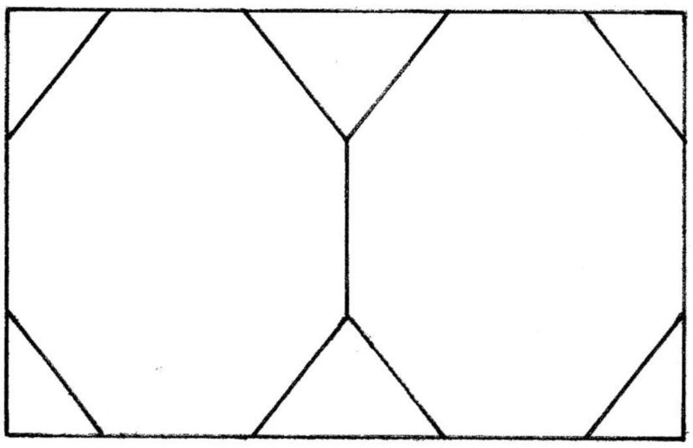

[t{4,3,3}]

Truncated
Hypercubic
Dodecahexaskew

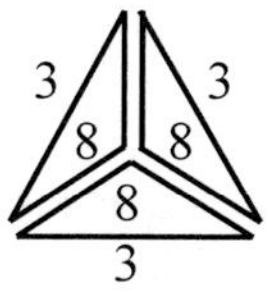

Fig.28: More Dodecahexaskews

EVEN MORE DODECAHEXASKEWS

Solids	V	E	F	S
tr octahedron	192	288	48 □	16
tr octahedron		^{2}96	64 ⬡	8
cube			64 ⬡	24
hexagon		(96) ‖	(96) □	32
Truncated Hyperoctahedral				
tr tetrahedron	48	^{2}96	64 ⬡	16
octahedron		^{4}24	64 △	8
4-lines			(96) ‖	24
hexagon			(96) ‖	32
Truncated Octahypertetrahedral				
tr tetrahedron	192	384	48 □	16
rhombicuboctahedron		^{2}96	64 △	8
cube			64 ⬡	24
hexaprism			96 □	32
		(96) ‖	(96) □	
Hyperhombicuboctetraprism				

Table 19: Even More Dodecahexaskews

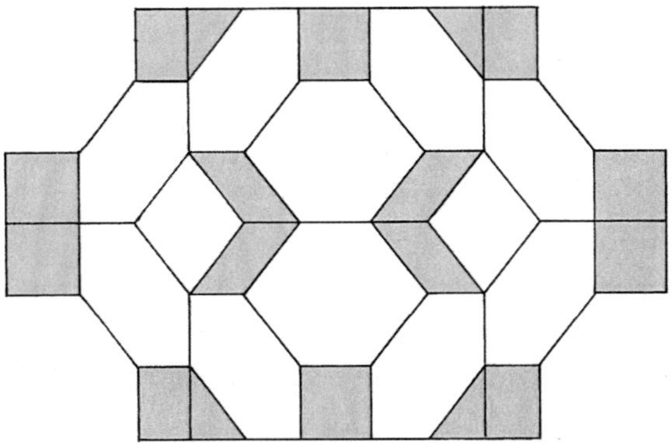

[tt^2{3,3,4}]

Truncated
Hyperoctahedral
Dodecahexaskew

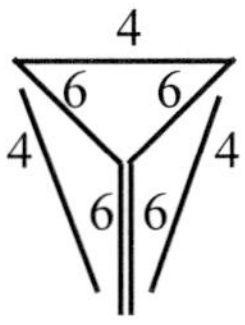

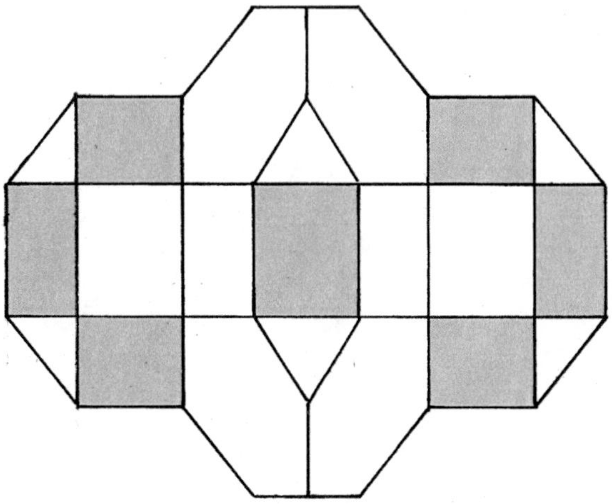

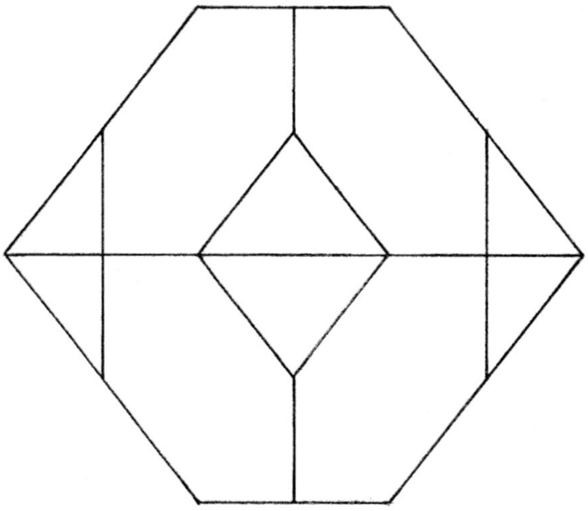

[o^2{4,3,3}]

Hyperhombi-
cuboctetraprism
Dodecahexaskew

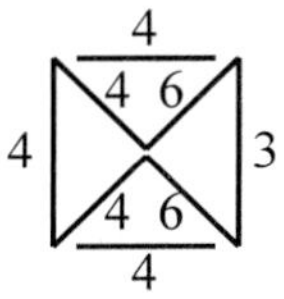

[t{3,3,4}]

Truncated
Octahypertetrahedral
Dodecahexaskew

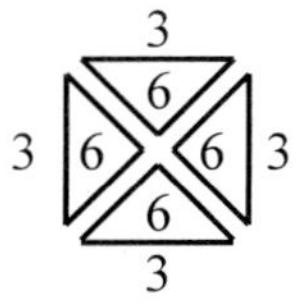

Fig.29: Even More Dodecahexaskews

MORE FAMILY GRIDS

Moving on down through the levels of complexity within the hypersolids, we now come to the simplest. Indeed it is the 'simplex' self dual hypertetrahedron {3,3,3} and its various truncations that form the basis of the next families of hyperskews to be considered. The first such family grid is shown adjoining in Figure 30, where the parent hyperskews are based on the two hypersolids t^3 and o^3{3,3,3}. These can be ascribed the amended Coxeter notations {6,4 |3 3} and {4,6 |3 3} respectively.

Again we find a central stretched version, comprising the faces of both parents along with added squares, based on the first odd of the family, o{3,3,3}. The two parents and their shared stretched version are shown opposite in Figure 31 and described numerically in Table 20.

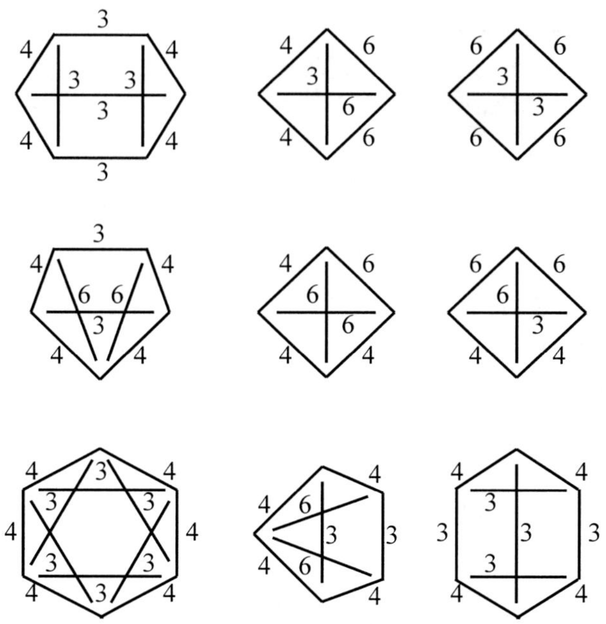

Fig.30: Tetratetraskew Grid [2-4,2-4]

The central member here comprises two equal spaces, each having five 4-connected truncated octahedra linked together by ten 2-connected hexaprisms, the latter's hexagon faces being the missing ones. The connectivity here and in the rest of this family is thus [2-4,2-4], leading to their naming as 'Tetratetraskews'. Again all of this family have a common Euler Number (V-E+F), this time -10, the smaller number very much in confirmation of the simpler arrangement of the surface between the two connected spaces.

Because of the symmetry inherent in the central vertex figure, the top left members of this family are the same as those at bottom right. Figure 32 and Table 21 on page 36 thus show only three further members of this family grid. The two intermediates are based on the tt^2 and o^2 {3,3,3} and the resultant in the corner is based on t^2t^2{3,3,3}.

TETRATETRASKEWS

Solids	V	E	F	S
tr tetrahedron	30	60	20 ⬡	5
tr tetrahedron			(10) △	5
{6,4 |3 3}			(10) △	
Hypertruncated Tetrahedral				
tr octahedron	120	240	20 ⬡	5
tr octahedron			90 □	5
hexaprism			(20) ⬡	10
hexaprism			(20) ⬡	10
nt^2t^2{4,6 |3 3}				
Truncated Hypertetratetraprism				
tetrahedron	20	60	30 □	5
tetrahedron			(20) △	5
triprism			(20) △	10
triprism				10
{4,6 |3 3}				
Hypertetraprism				

Table 20: Tetratetraskews

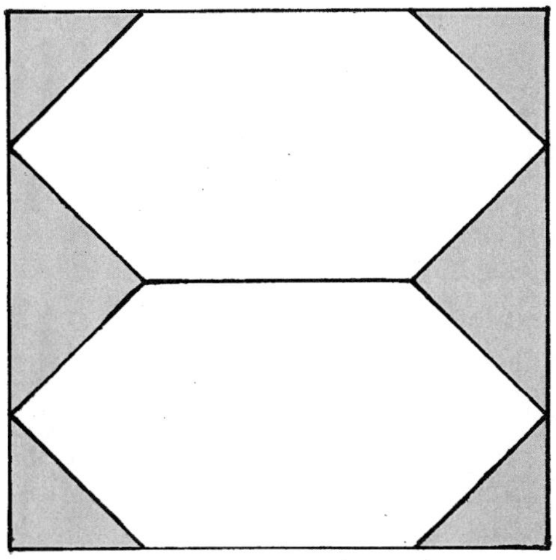

{6,4 |3 3} [t^3{3,3,3}]
Hypertruncated
Tetrahedral
Tetratetraskew

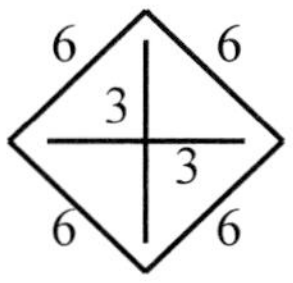

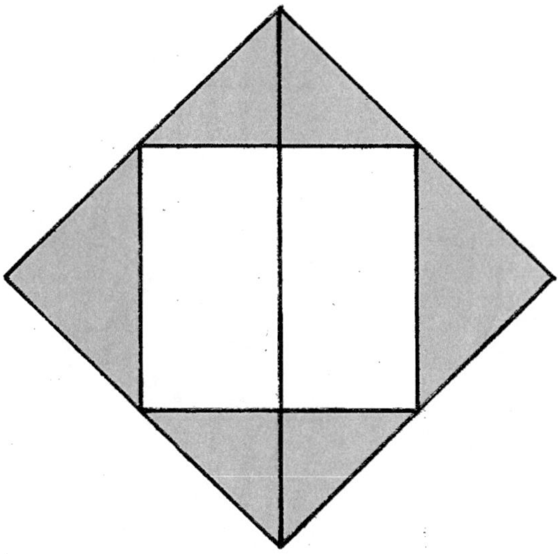

{4,6 |3 3} [o^3{3,3,3}]

Hypertetraprism
Tetratetraskew

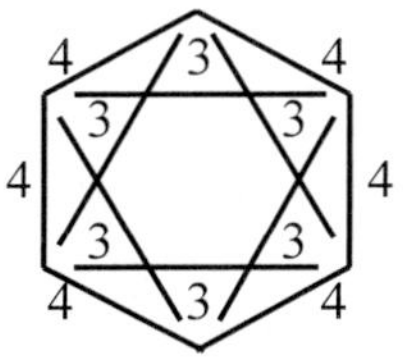

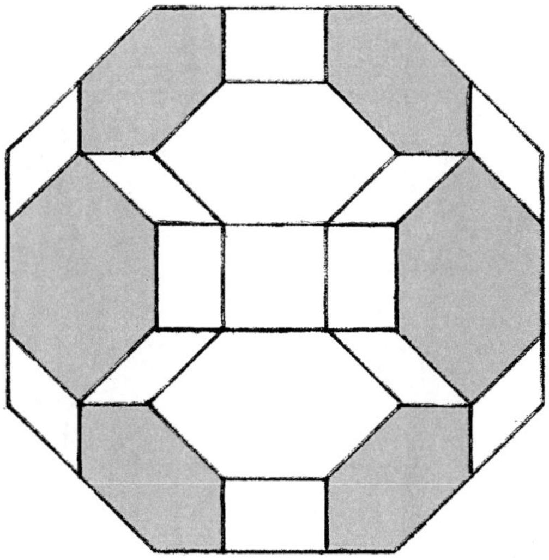

nt^2t^2{4,6 |3 3} [o{3,3,3}]
Truncated
Hypertetratetraprism
Tetratetraskew

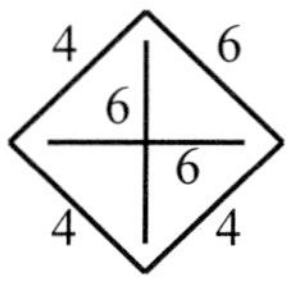

Fig.31: Tetratetraskews

MORE TETRATETRASKEWS

Solids	V	E	F	S
tr tetrahedron	60	120	20 ⬡	5
tr octahedron			30 □	5
triprism			(20) △	10
			(10) ⬡	
Truncated Hypertetratetrahedral				
octahedron	30	90	20 △	5
cuboctahedron			30 □	5
triprism			(20) △	10
			(10) △	
Hyperhombitetratetrahedral				
tr tetrahedron	60	150	20 △	5
cuboctahedron			60 □	5
triprism			(20) ⬡	10
hexaprism			(20) △	10
Hyperhombitetratetraprism				

Table 21: More Tetratetraskews

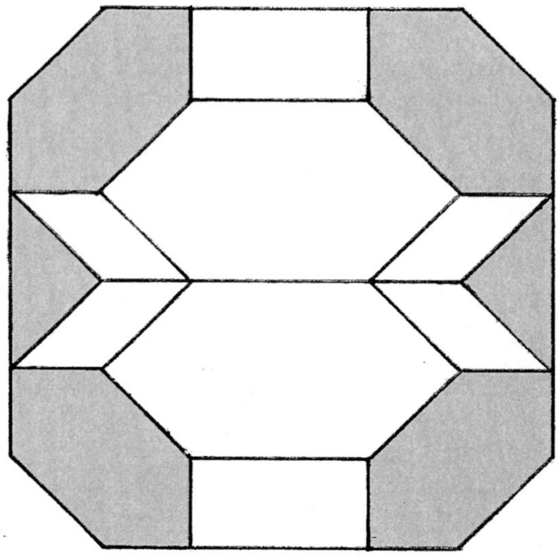

$[tt^2\{3,3,3\}]$

Truncated Hyper-
tetratetrahedral
Tetratetraskew

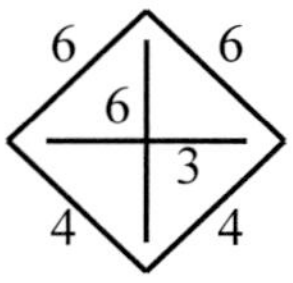

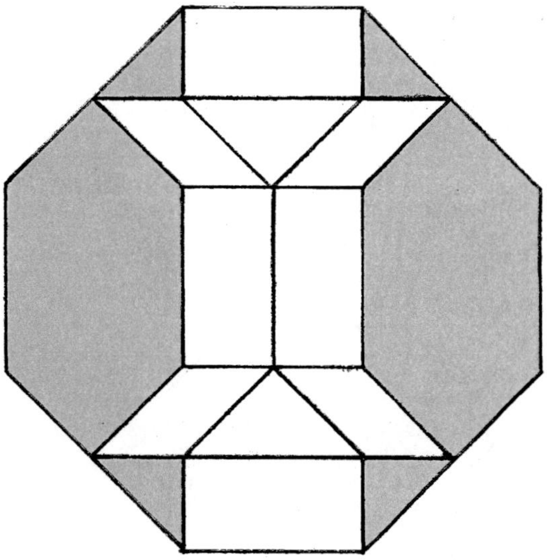

$[o^2\{3,3,3\}]$

Hyperhombi-
tetratetraprism
Tetratetraskew

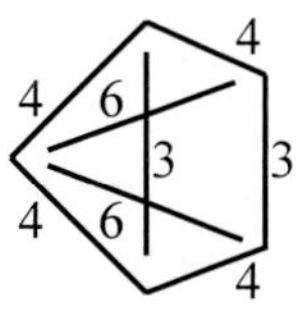

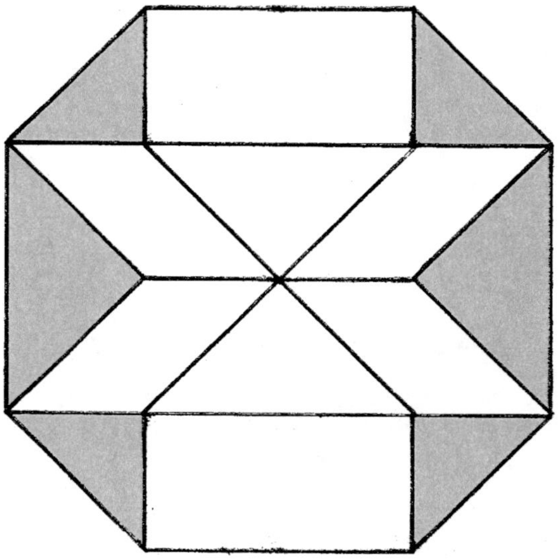

$[t^2t^2\{3,3,3\}]$

Hyperhombi-
tetratetrahedral
Tetratetraskew

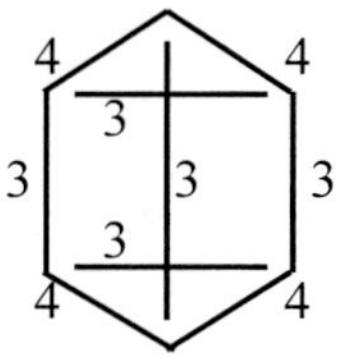

Fig.32: More Tetratetraskews

TETRATRISKEW GRID

The next family grid we will examine is still based on the hypertetrahedron {3,3,3} and its various truncations, but this time the central tetrahedral vertex figure of o{3,3,3} is aligned to have a square and a hexagon as the missing faces.

Figure 33 adjoining shows the family grid produced with this as central member, which again has some repeats such that in total there are only five different hyperskews produced: one central, two intermediates and two resultants.

One of the resultants, appearing at top left and bottom right, is self dual with six hexagons at each vertex, and can be given the amended Coxeter notation {6,6 |3 2}. Its faces are those of the t{3,3,3} truncated hypertetrahedron, with triangles and digons missing.

Figure 34 overleaf shows this at top right with below it the central member of this grid, which comprises one space with truncated octahedra, 4-connected by missing hexagon faces, and the other space hexaprisms, 3-connected by missing square faces. Table 22 gives the relatively simple numerical data for these two.

The other three members of this family, two intermediates and a resultant are shown on page 39 in Figure 35, their numerical properties appearing in the adjoining Table 23.

This family thus has connectivity [3-3,4-4] and its members are according dubbed 'Tetratriskews'. They too have a common Euler Number (V-E+F) equal to -20.

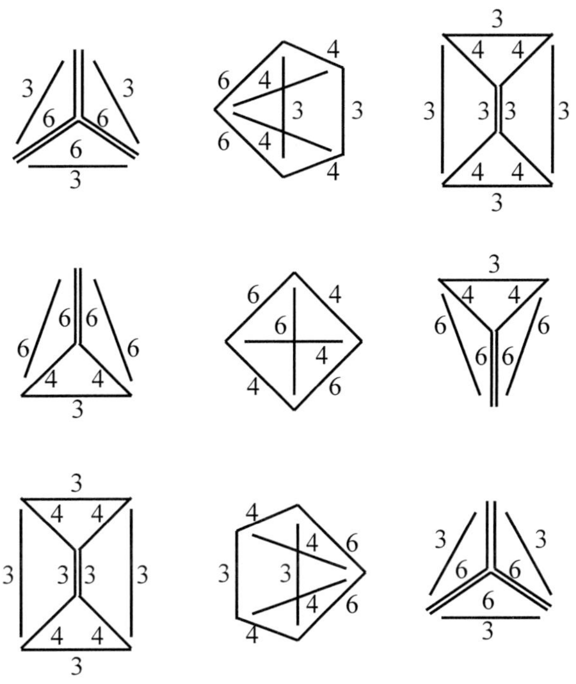

Fig.33: Tetratriskew Grid [3-3,4-4]

RHOMBIC TETRATRISKEW

Because {6,6 |3 2} is self dual, there is a rhombic form that shares the same [3-3,4-4] connectivities of this family. A necessarily small section of this Rhombic Tetratriskew is shown overleaf at bottom left of Figure 34, essentially the dual of the central member of the grid, based on o{3,3,3}.

The rhomb faces, which meet 4-fold at some vertices and 6-fold at others, are those of the Rhombic Dodecahedron with the diagonals in proportion with the vertex edge lengths of a square and of a hexagon, i.e. $\sqrt{2}:\sqrt{3}$ or 1:1.22474, giving vertex angles of 78.46° and 101.54°.

TETRATRISKEWS

Solids	V	E	F	S
hexagon	20	30	20 ⬡	10
tr tetrahedron		^{3}10	(20) △	5
tetrahedron			(30) ‖	5
3-lines				10
Truncated Hypertetrahedral			{6,6 |3 2}	
hexaprism	120	240	20 ⬡	10
tr octahedron			20 ⬡	5
tr octahedron			60 □	5
hexaprism			(20) ⬡	10
nt^2t^2{6,6 |3 2}			(30) □	
Truncated Hypertetratetraprism				
rhomb	100	240	120 ◇	n/a
Rhombic				

Table 22: Tetratriskews

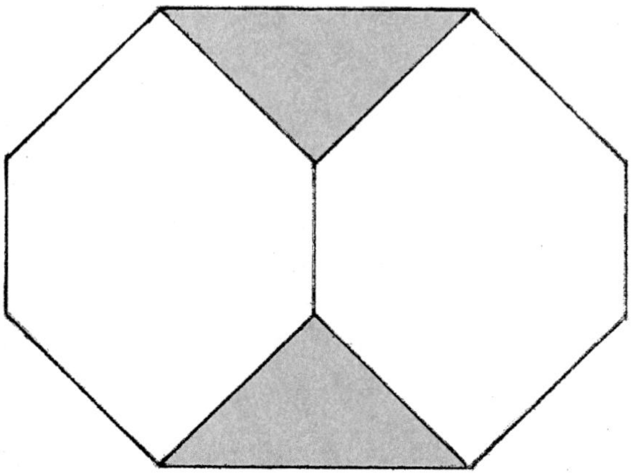

{6,6 |3 2} [t{3,3,3}]
Truncated
Hypertetrahedral
Tetratriskew

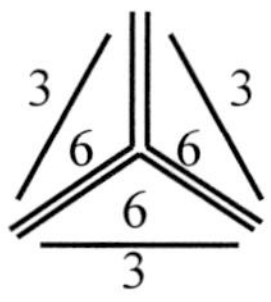

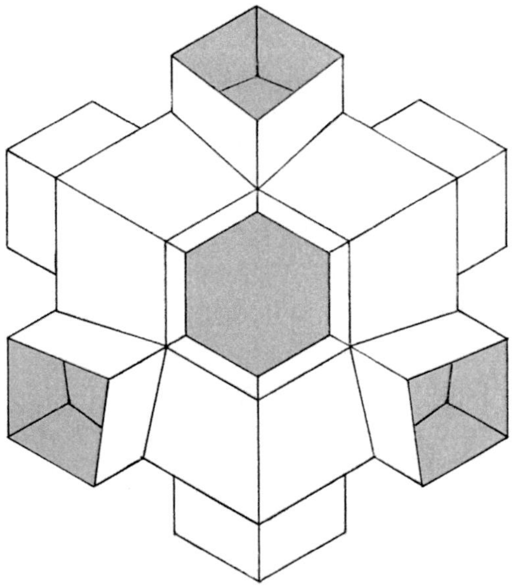

Rhombic
Tetratriskew

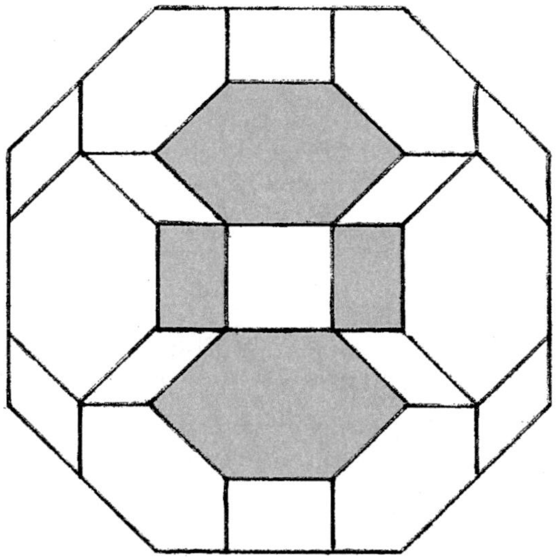

nt^2t^2{6,6 |3 2} [o{3,3,3}]
Truncated
Hypertetratetraprism
Tetratriskew

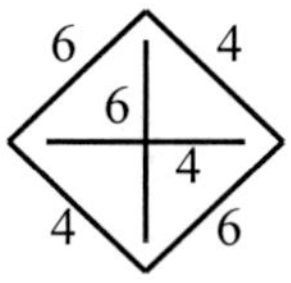

Fig.34: Tetratriskews

MORE TETRATRISKEWS

Solids	V	E	F	S
hexagon	60	90	20 ⬡	10
tr octahedron		$^{2}30$	20 △	5
tr tetrahedron			30 □	5
triprism		(30) ‖	(20) ⬡	10
Truncated Hypertetratetrahedral				
triangle	30	60	20 △	10
cuboctahedron		$^{2}30$	20 △	5
octahedron			30 □	5
triprism		(30) ‖	(20) △	10
Hyperhombitetratetrahedral				
triprism	60	150	20 △	10
cuboctahedron			20 ⬡	5
tr tetrahedron			30 □	5
hexaprism			(20) △	10
			(30) □	
Hyperhombitetratetraprism				

Table 23: More Tetratriskews

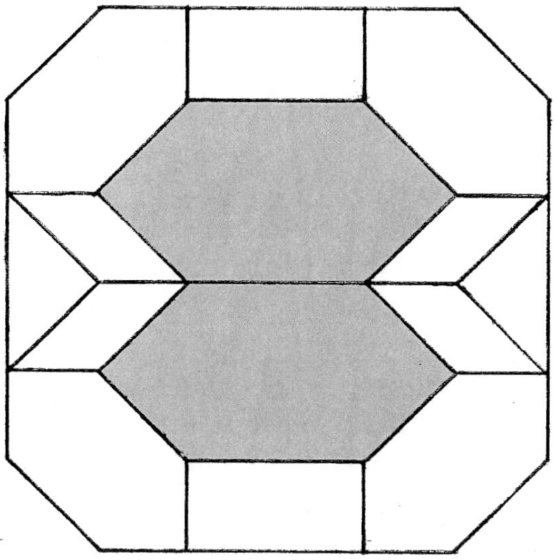

[tt²{3,3,3}]

Truncated Hypertetratetrahedral Tetratriskew

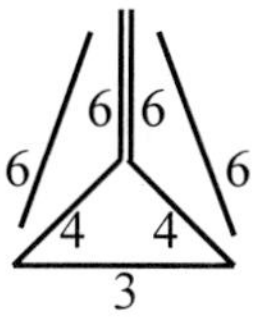

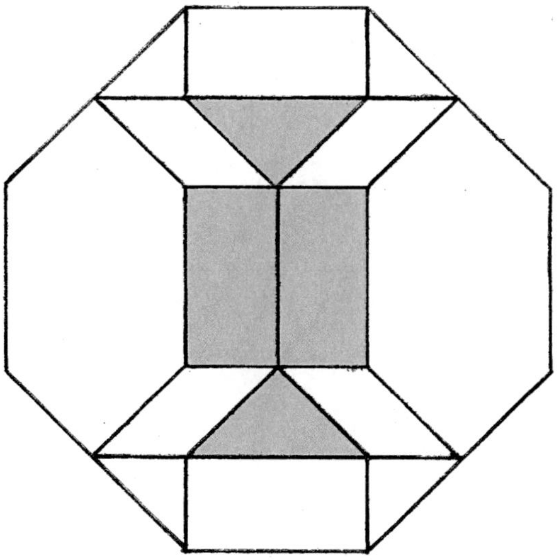

[o²{3,3,3}]

Hyperhombi-tetratetraprism Tetratriskew

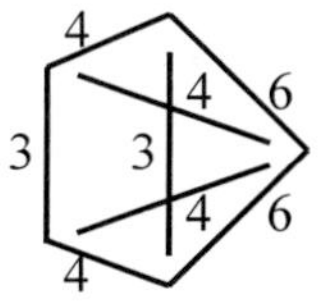

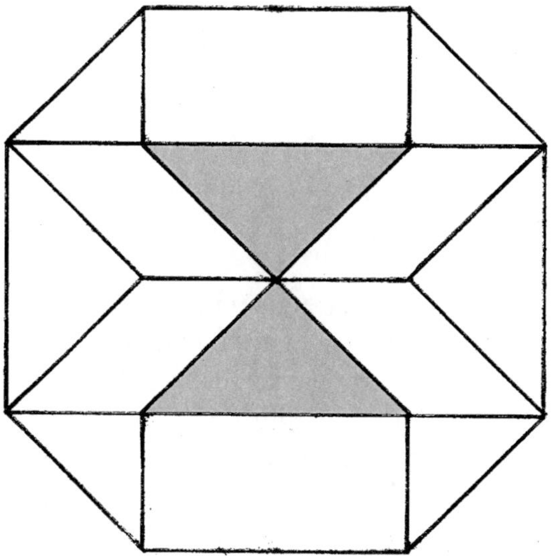

[t²t²{3,3,3}]

Hyperhombi-tetratetrahedral Tetratriskew

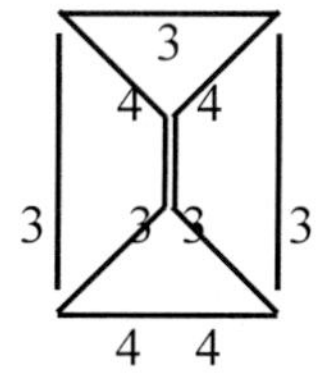

Fig.35: More Tetratriskews

HEXAHEXASKEW GRID

Following on from the previous two families we can find the final alignment of this o{3,3,3} central vertex figure, that with two squares as the missing faces.

The central member here once again comprises two equal spaces, each with truncated octahedra 6-connected by their missing squares to 3-connected hexaprisms.

These we will call ‘Hexahexaskews’, the full family grid for which is shown adjoining in Figure 36. Again we find that there are repeats, this time such that there are only six different members to this family. They all share the same Euler Number -30, as before the sum of those of the two preceding families based on the hypersolid o{3,3,3} as central member.

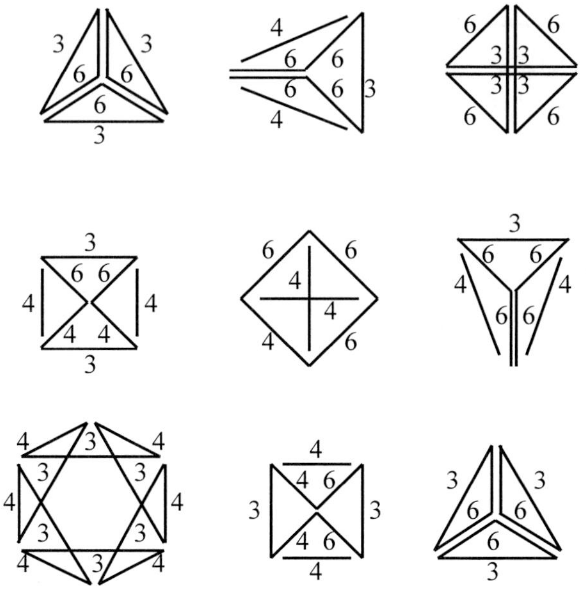

Fig.36: Hexahexaskew Grid [3-6,3-6]

The central member and two resultants based on t^3 and o^3{3,3,3} are shown opposite in Figure 37, their numerical properties being summarised in Table 24.

Figure 38 and Table 25 overleaf on page 42 show the remainder of this family, two intermediates and a resultant, based on tt^2 and o^2{3,3,3} and t{3,3,3} respectively.

Our previous experience with ‘incomplete’ family grids tells us that the two new tetrahedral vertex figures thrown up by Figure 30 will lead no further. The intermediate versions at top centre and right centre will only produce an incomplete grid including the already described members of the row or column they sit in. The resultant at top right has two odd valued diagonals, both {3} and not available as a truncation of any polygon.

HEXAHEXASKEWS

Solids	V	E	F	S
tr tetrahedron	30	^{2}60	20 △	5
tr tetrahedron			20 △	5
triangle			20 ⬡	10
triangle			(60) ‖	10
Hypertruncated Tetrahedral				
tr octahedron	120	240	20 ⬡	5
tr octahedron			40 ⬡	5
hexaprism			30 □	10
hexaprism			(60) □	10
Truncated Hypertetratetraprism				
tetrahedron	20	^{2}60	20 △	5
tetrahedron			20 △	5
triprism			30 □	10
triprism			(60) ‖	10
Hypertetraprism				

Table 24: Hexahexaskews

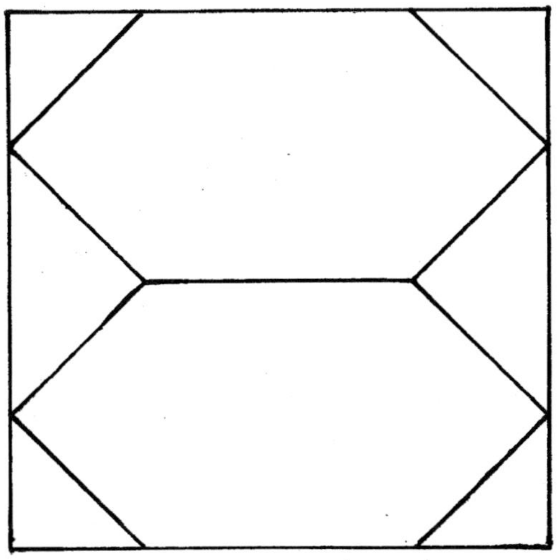

$[t^3\{3,3,3\}]$

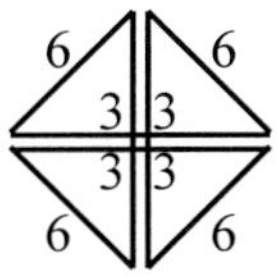

Hypertruncated
Tetrahedral
Hexahexaskew

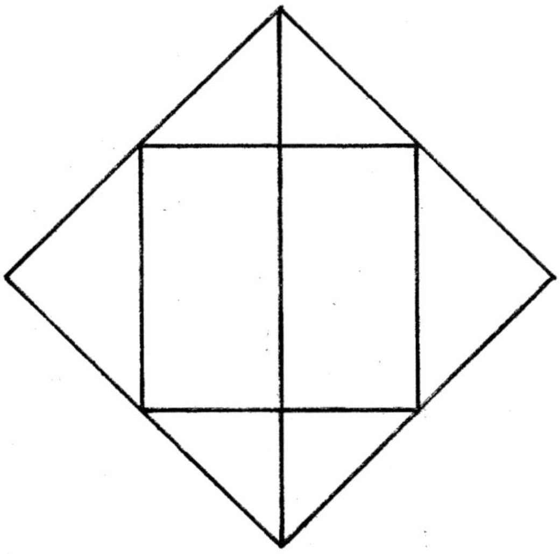

$[o^3\{3,3,3\}]$

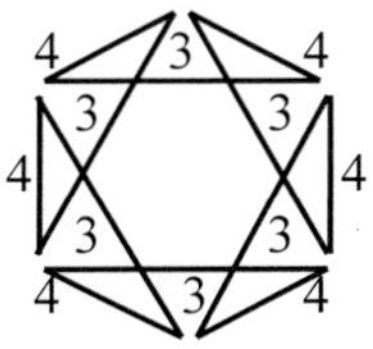

Hypertetraprism
Hexahexaskew

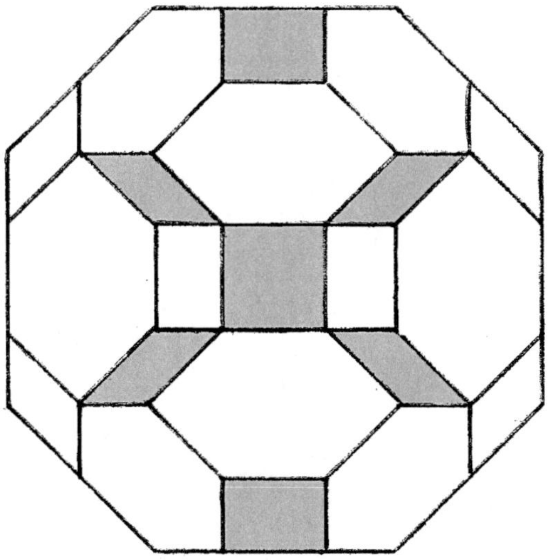

$[o\{3,3,3\}]$

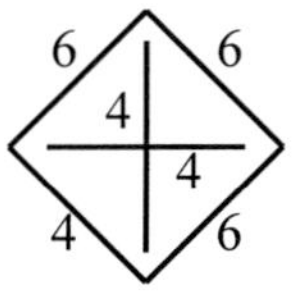

Truncated
Hypertetratetraprism
Hexahexaskew

Fig.37: Hexahexaskews

MORE HEXAHEXASKEWS

Solids	V	E	F	S
tr octahedron	60	90	20 △	5
tr tetrahedron		$^{2}30$	20 ⬡	5
triprism			20 ⬡	10
hexagon		(30) ∥	(30) □	10
Truncated Hypertetratetrahedral				
tr tetrahedron	20	$^{2}30$	20 ⬡	5
tetrahedron		$^{3}10$	20 △	5
3-lines			(30) ∥	10
hexagon			(30) ∥	10
Truncated Hypertetrahedral				
tr tetrahedron	60	120	20 △	5
cuboctahedron		$^{2}30$	20 ⬡	5
triprism			20 △	10
hexaprism			30 □	10
		(30) ∥	(30) □	
Hyperhombitetratetraprism				

Table 25: More Hexahexaskews

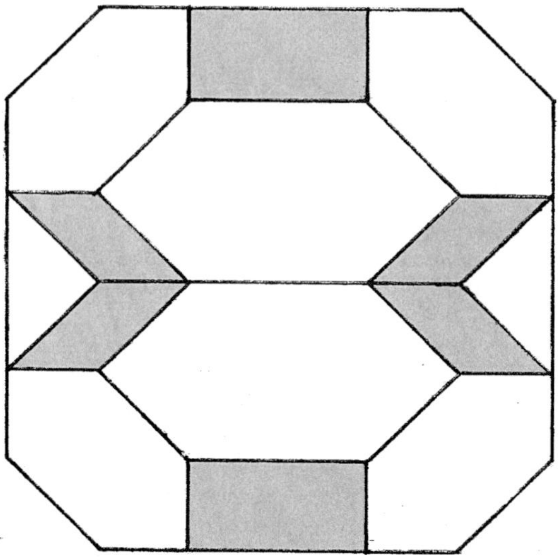

[tt^{2}{3,3,3}]

Truncated
Hypertetratetrahedral
Hexahexaskew

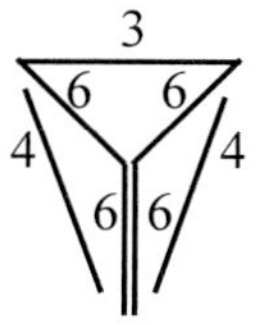

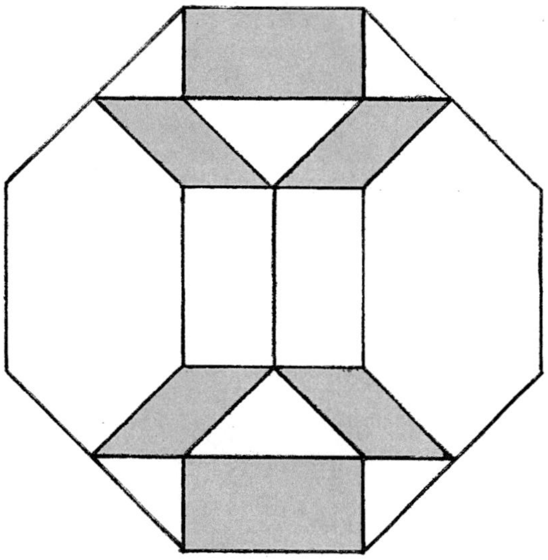

[o^{2}{3,3,3}]

Hyperhombi-
tetratetraprism
Hexahexaskew

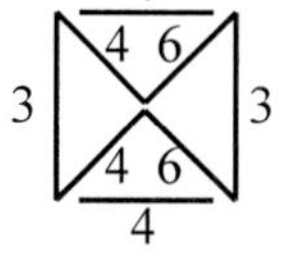

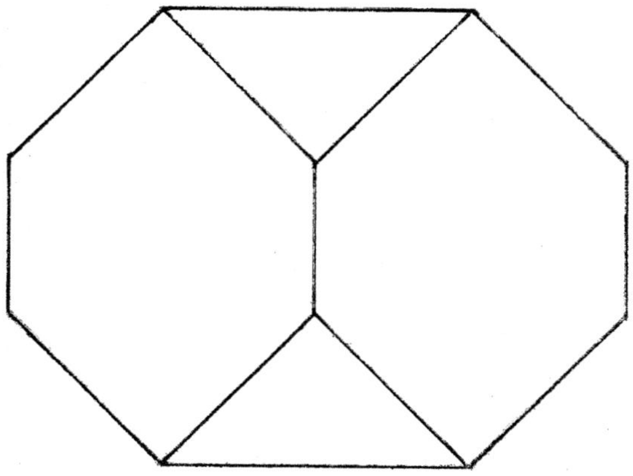

[t{3,3,3}]

Truncated
Hypertetrahedral
Hexahexaskew

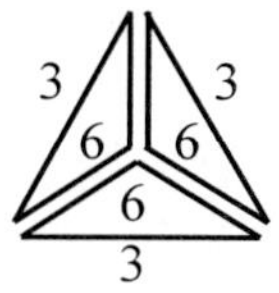

Fig.38: More Hexahexaskews

EVEN MORE FAMILY GRIDS

Having moved right down through the levels of complexity within the hypersolids, we have now seen the 'simplex' self dual hypertetrahedron {3,3,3} and its various truncations that formed the basis of the three families of hyperskews just examined.

There remains one further hypersolid of medium complexity, the hyperoctahedron {3,4,3}, another self dual, which with its truncations forms the basis of our last three families of hyperskews.

The first family grid derived from this is shown adjoining in Figure 39, where the parent hyperskews, top right and bottom left, are based on the t^3 and o^3{3,4,3}. These can be ascribed the amended Coxeter notations {8,4 |3 3} and {4,8 |3 3} respectively.

The first of these is a simple surface of 144 octagon faces separating two equal spaces each comprising 24 truncated cubes linked by missing triangle faces.

Again we find a central stretched version, comprising the faces of both parents along with added squares, based on the first odd of the family, o{3,4,3}. The two parents and their shared stretched version are shown overleaf in Figure 40 and described numerically in Table 26.

The central member here also comprises two equal spaces, each having twenty four 8-connected truncated cuboctahedra linked together by ninety six 2-connected hexaprisms, the latter's hexagon faces being the missing ones. The connectivity here and in the rest of this family is thus [2-8,2-8], leading to their naming as 'Octoctaskews'.

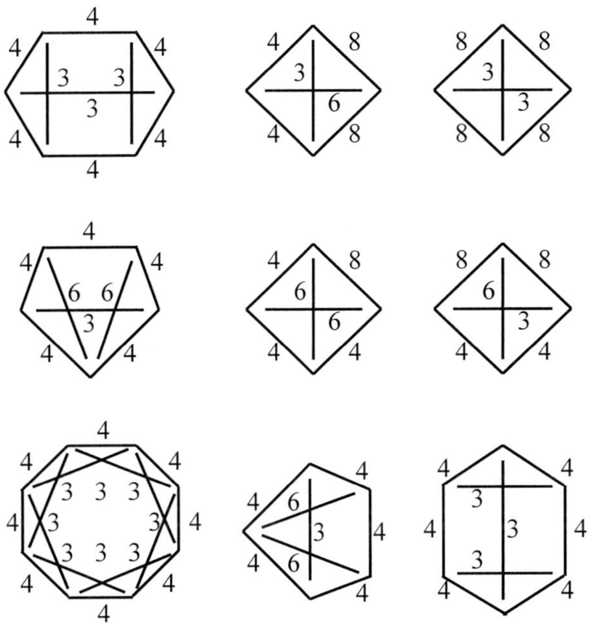

Fig.39: Octoctaskew Grid [2-8,2-8]

The Euler Number (V-E+F) for this whole family is -144, its larger size indicative of greater complexity than we saw in the last three families based on the hypertetrahedron {3,3,3} hypersolids.

Because of the symmetry inherent in the central vertex figure, the top left members of this family are the same as those at bottom right. Figure 41 and Table 27 on page 45 thus show only three further members of this family grid. The two intermediates are based on the tt^2 and o^2 {3,4,3} and the resultant in the corner is based on t^2t^2{3,4,3}.

OCTOCTASKEWS

Solids	V	E	F	S
tr cube	288	576	144 ⯃	24
tr cube			(96) △	24
{8,4 |3 3}			(96) △	
Hypertruncated Cubic				
tr cuboctahedron	1152	2304	144 ⯃	24
tr cuboctahedron			864 □	24
hexaprism			(192) ⬡	96
hexaprism			(192) ⬡	96
nt^2t^2{4,8 |3 3}				
Truncated Hypercuboctaprism				
octahedron	144	576	288 □	24
octahedron			(192) △	24
triprism			(192) △	96
triprism				96
{4,8 |3 3}				
Hyperoctaprism				

Table 26: Octoctaskews

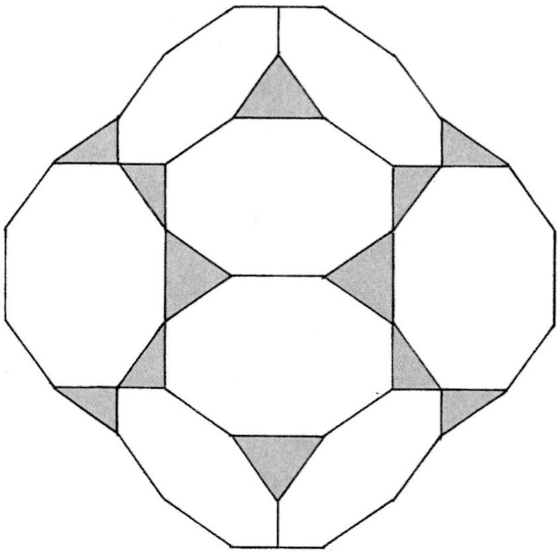

{8,4 |3 3} [t^3{3,4,3}]
Hypertruncated
Cubic
Octoctaskew

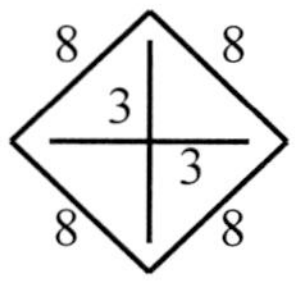

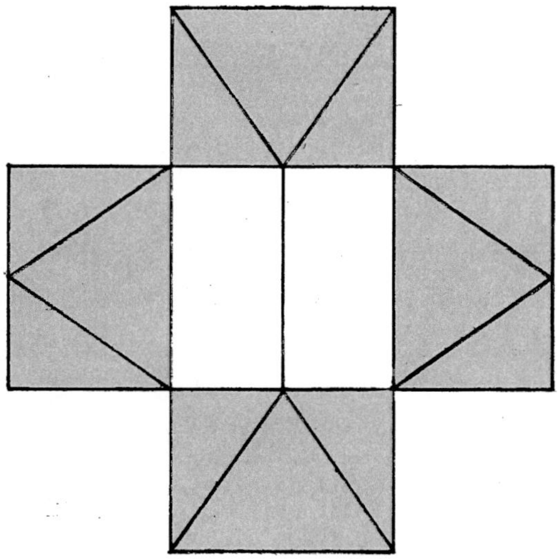

{4,8 |3 3} [o^3{3,4,3}]

Hyperoctaprism
Octoctaskew

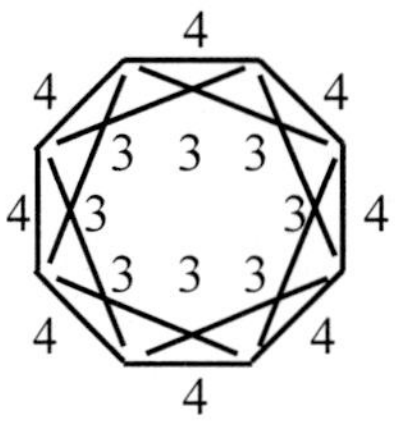

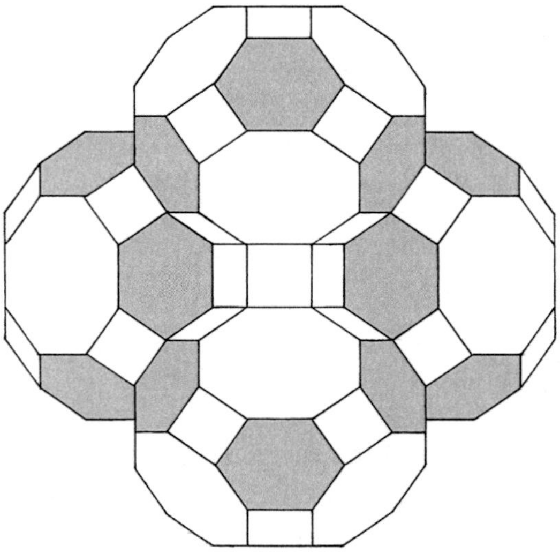

nt^2t^2{4,8 |3 3} [o{3,4,3}]
Truncated
Hypercuboctaprism
Octoctaskew

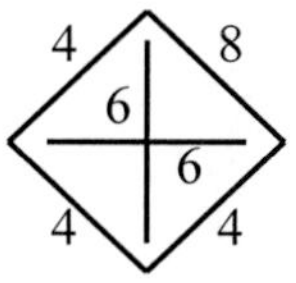

Fig.40: Octoctaskews

MORE OCTOCTASKEWS

Solids	V	E	F	S
tr cube	576	1152	144 octagon	24
tr cuboctahedron			288 □	24
triprism			(192) △	96
			(96) hexagon	
Truncated Hypercuboctahedral				
cuboctahedron	288	864	144 □	24
rhombicuboctahedron			288 □	24
triprism			(192) △	96
			(96) △	
Hyperhombicuboctahedral				
tr octahedron	576	1440	144 □	24
rhombicuboctahedron			576 □	24
triprism			(192) hexagon	96
hexaprism			(192) △	96
Hyperhombicuboctaprism				

Table 27: More Octoctaskews

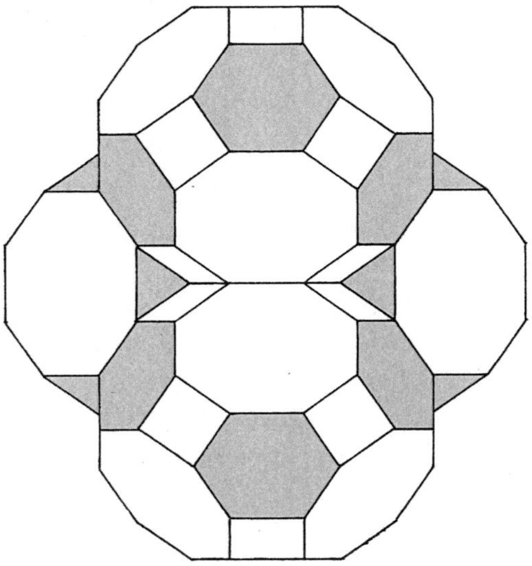

[tt^2{3,4,3}]

Truncated Hyper-cuboctahedral Octoctaskew

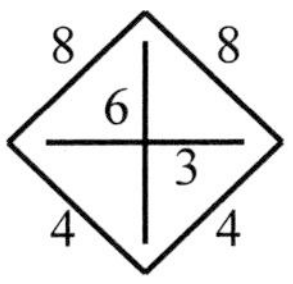

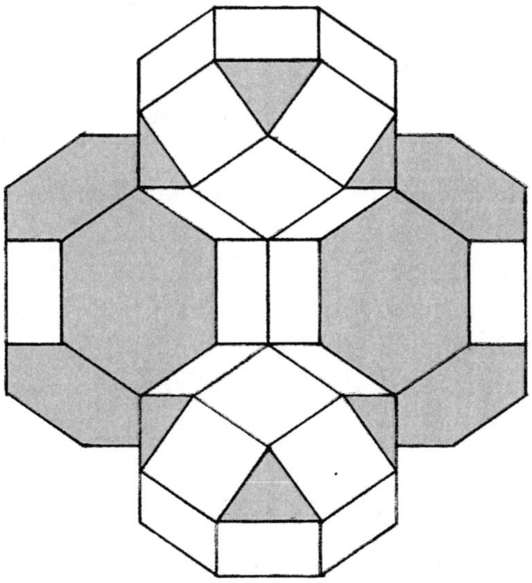

[o^2{3,4,3}]

Hyperhombi-cuboctaprism Octoctaskew

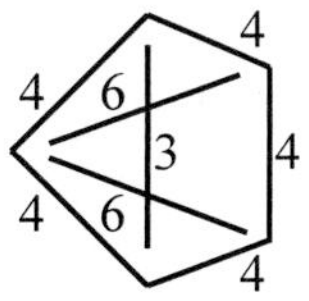

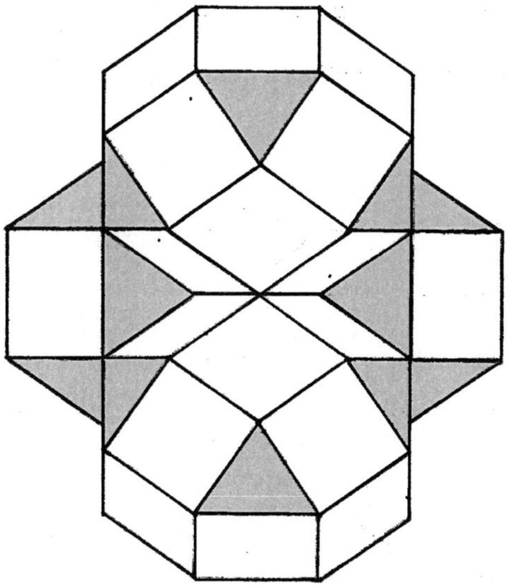

[t^{2}t^2{3,4,3}]

Hyperhombi-cuboctahedral Octoctaskew

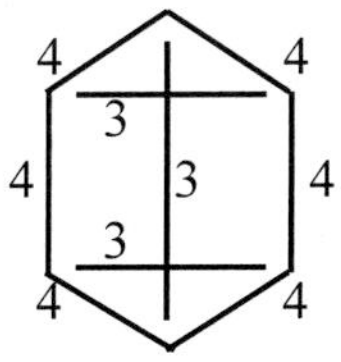

Fig.41: More Octoctaskews

HEXATRISKEW GRID

The next family grid we will examine is still based on the hyperoctahedron {3,4,3} and its various truncations, but this time the central tetrahedral vertex figure of o{3,4,3} is aligned to have a square and an octagon as the missing faces.

Figure 42 adjoining shows the family grid produced with this as central member, which again has some repeats such that in total there are only five different hyperskews produced: one central, two intermediates and two resultants.

One of the resultants, appearing at top left and bottom right, is self dual with six hexagons at each vertex, and can be given the amended Coxeter notation {6,6 |4 2}. Its faces are those of the t{3,4,3} truncated hyperoctahedron, with squares and digons missing.

Figure 43 opposite shows this at top right with below it the central member of this grid, which comprises one space with truncated cuboctahedra, 6-connected by missing octagon faces, and the other space hexaprisms 3-connected by missing square faces. Table 28 gives the relevant numerical data for these.

The other three members of this family, two intermediates and a resultant are shown on page 48 in Figure 44, their numerical properties appearing in the adjoining Table 29.

This family thus has connectivity [3-3,6-6] and its members are according dubbed 'Hexatriskews'. The Euler Number here once account is taken of multiple faces and edges is -192.

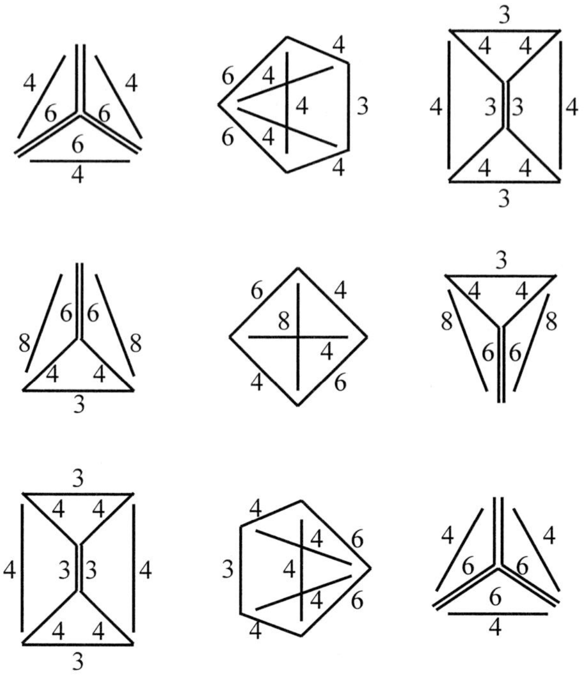

Fig.42: Hexatriskew Grid [3-3,6-6]

RHOMBIC HEXATRISKEW

Because {6,6 |4 2} is self dual, there is a rhombic form that shares the same [3-3,6-6] connectivities of this family. A small section of this Rhombic Hexatriskew is shown opposite at bottom left of Figure 43, essentially the dual of the central member of the grid, based on o{3,4,3}.

The rhomb faces, which meet 4-fold at some vertices and 6-fold at others, are again those of the Rhombic Dodecahedron with the diagonals in proportion with the vertex edge lengths of a square and of a hexagon, i.e. √2:√3 or 1:1.22474, giving vertex angles of 78.46° and 101.54°.

HEXATRISKEWS

Solids	V	E	F	S
hexagon	192	288	192 ⬡	96
tr octahedron		^{3}96	(144) □	24
cube			(288) ‖	24
3-lines				96
Truncated Hyperoctahedral			{6,6 |4 2}	
hexaprism	1152	2304	192 ⬡	96
tr cuboctahedron			192 ⬡	24
tr cuboctahedron			576 □	24
hexaprism			(144) ⯃	96
nt^2t^2{6,6 |4 2}			(288) □	
Truncated Hypercuboctaprism				
rhomb	960	2304	1152 ◇	n/a
Rhombic				

Table 28: Hexatriskews

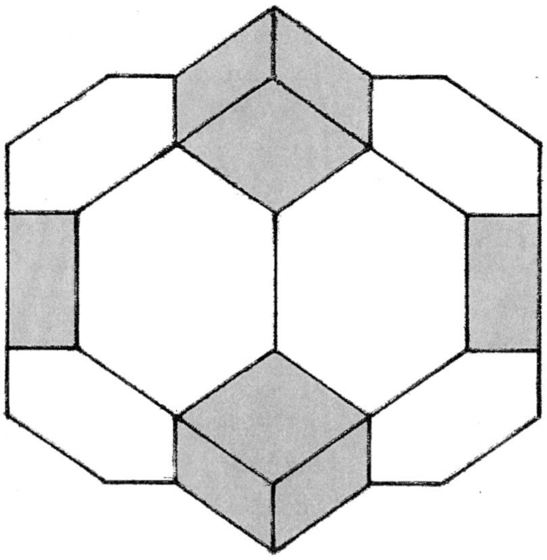

{6,6 |4 2} [t{3,4,3}]
Truncated
Hyperoctahedral
Hexatriskew

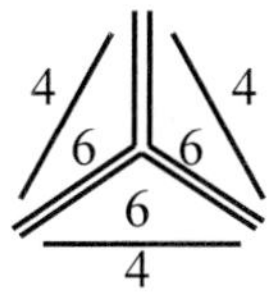

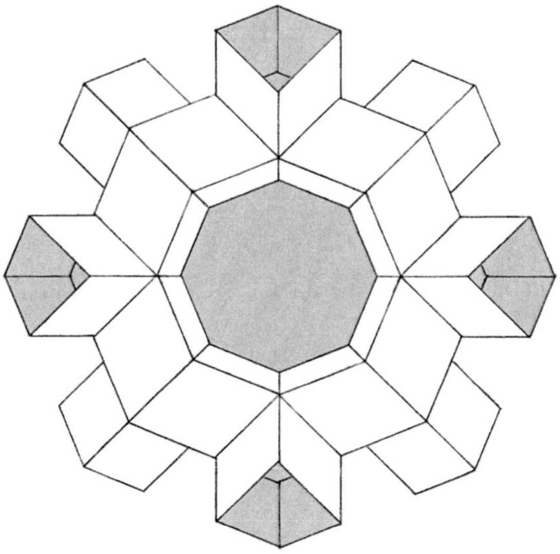

Rhombic
Hexatriskew

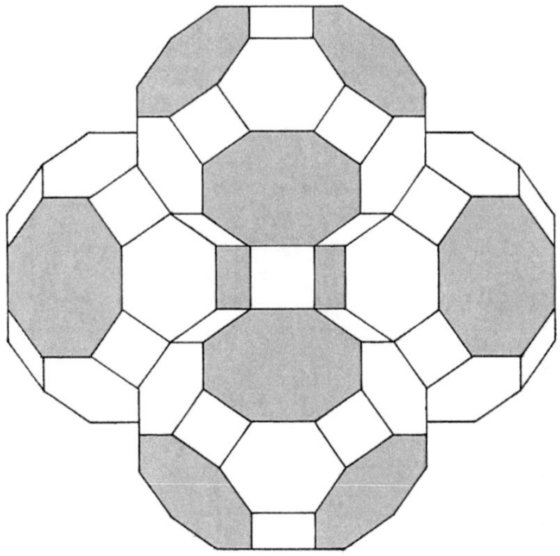

nt^2t^2{6,6 |4 2} [o{3,4,3}]
Truncated
Hypercuboctaprism
Hexatriskew

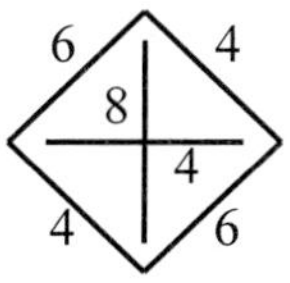

Fig.43: Hexatriskews

MORE HEXATRISKEWS

Solids	V	E	F	S
hexagon	576	864	192 ⬡	96
tr cuboctahedron		^{2}288	192 △	24
tr cube			288 □	24
triprism		(288) ‖	(144) ⯃	96
Truncated Hypercuboctahedral				
triangle	288	576	192 △	96
rhombicuboctahedron		^{2}288	192 △	24
cuboctahedron			288 □	24
triprism		(288) ‖	(144) □	96
Hyperhombicuboctahedral				
triprism	576	1440	192 △	96
rhombicuboctahedron			192 ⬡	24
tr octahedron			288 □	24
hexaprism			(144) □	96
			(288) □	
Hyperhombicuboctaprism				

Table 29: More Hexatriskews

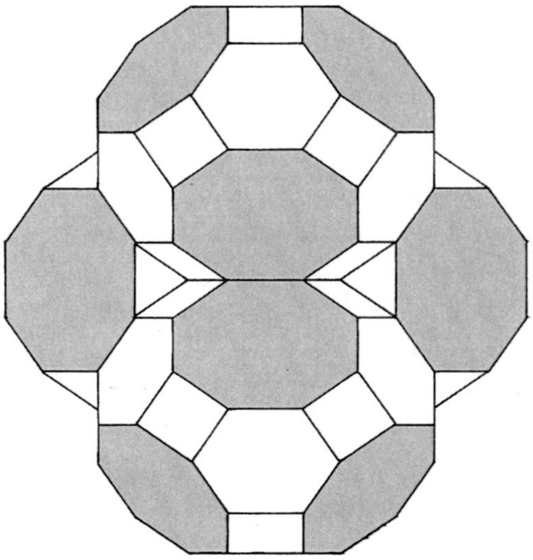

[tt^2{3,4,3}]

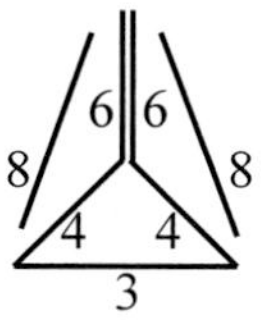

Truncated Hypercuboctahedral Hexatriskew

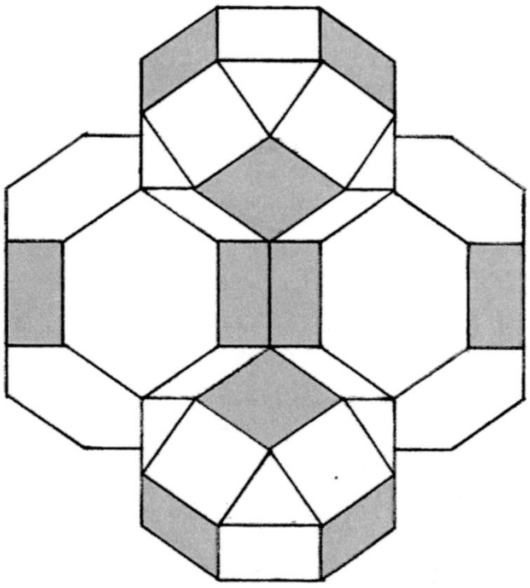

[o^2{3,4,3}]

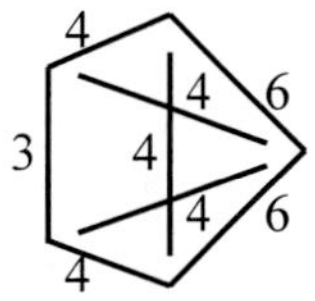

Hyperhombi-cuboctaprism Hexatriskew

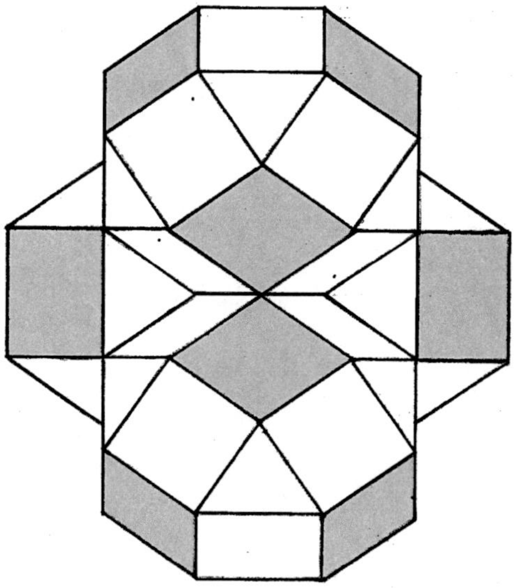

[t^{2}t^2{3,4,3}]

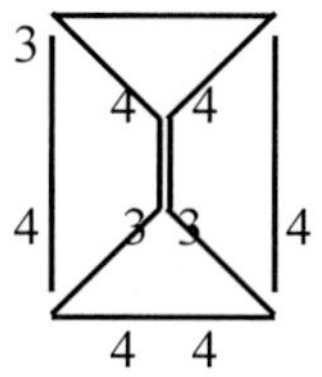

Hyperhombi-cuboctahedral Hexatriskew

Fig.44: More Hexatriskews

DODECADODECASKEW GRID

Following on from the previous two families we can find the final alignment of this o{3,4,3} central vertex figure, that with two squares as the missing faces.

The central member here once again comprises two equal spaces, each with truncated cuboctahedra 12-connected by their missing squares to 3-connected hexaprisms.

These we will call 'Dodecadodecaskews', the full family grid for which is shown adjoining in Figure 45. Again we find that there are repeats, this time such that there are only six different members to this family. As might be expected, once multiple values are taken into account they all share the same Euler Number, this time -336, the sum of the previous two family's numbers.

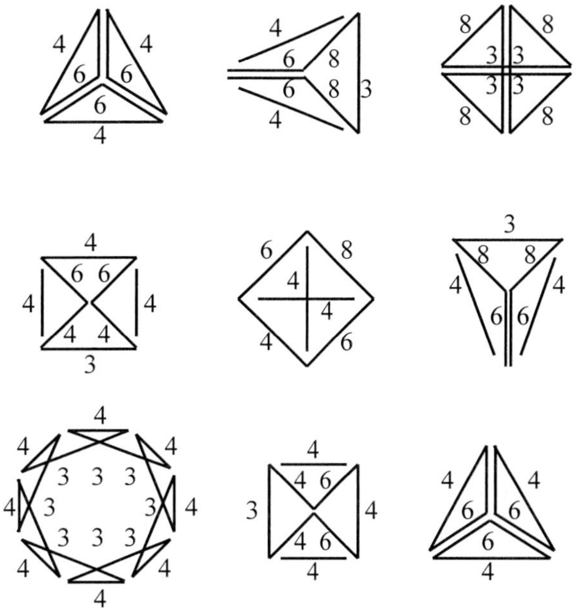

Fig.45:Dodecadodecaskew Grid [3-12,3-12]

The central member and two resultants based on t^3 and $o^3\{3,4,3\}$ are shown overleaf in Figure 46, their numerical properties being summarised in Table 30.

Figure 47 and Table 31 on page 51 show the remainder of this family, two intermediates and a resultant, based on tt^2 and $o^2\{3,4,3\}$ and t{3,4,3} respectively.

Our previous experience with 'incomplete' family grids tells us that the two new tetrahedral vertex figures thrown up by Figure 39 will as before lead no further. The intermediate versions at top centre and right centre will only produce an incomplete grid including the already described members of the row or column they sit in. The resultant at top right has two odd valued diagonals, both {3} and not a truncation of any polygon.

DODECADODECASKEWS

Solids	V	E	F		S
truncated cube	288	^{2}576	144	⯃	24
truncated cube			192	△	24
triangle			192	△	96
triangle			(576)	‖	96
Hypertruncated Cubic					
tr cuboctahedron	1152	2304	144	⯃	24
tr cuboctahedron			384	⬡	24
hexaprism			288	□	96
hexaprism			(576)	□	96
Truncated Hypercuboctaprism					
octahedron	144	^{2}576	192	△	24
octahedron			192	△	24
triprism			288	□	96
triprism			(576)	‖	96
Hyperoctaprism					

Table 30: Dodecadodecaskews

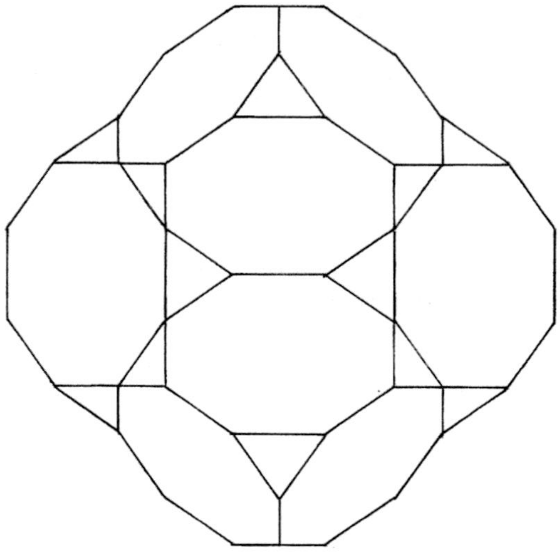

[t^3{3,4,3}]

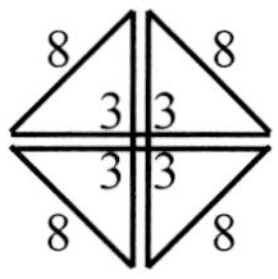

Hypertruncated
Cubic
Dodecadodecaskew

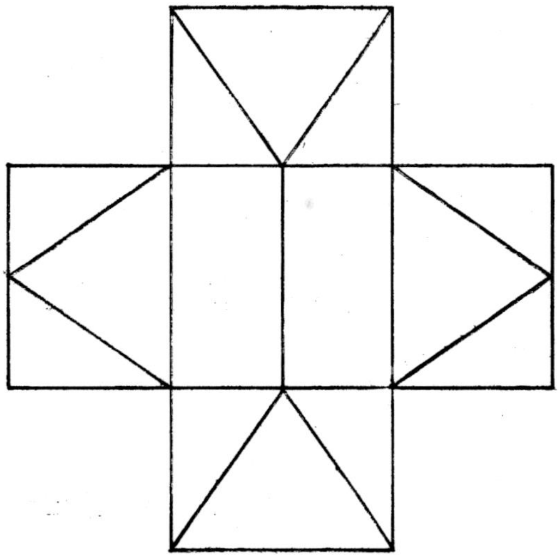

[o^3{3,4,3}]

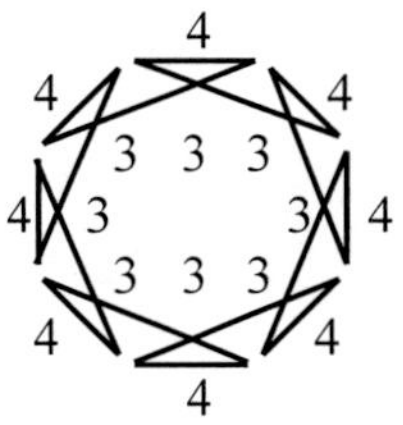

Hyperoctaprism
Dodecadodecaskew

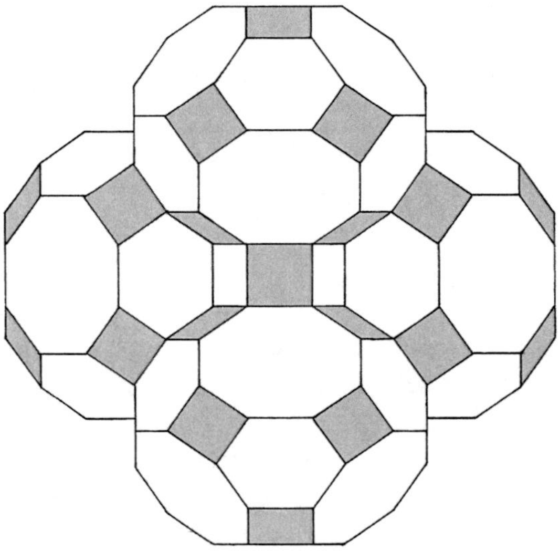

[o{3,4,3}]

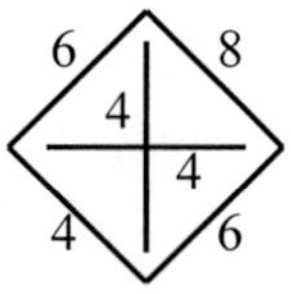

Truncated
Hypercuboctaprism
Dodecadodecaskew

Fig.46: Dodecadodecaskews

MORE DODECADODECASKEWS

Solids	V	E	F	S
tr cuboctahedron	576	864	192 ⬡	24
truncated cube		^{2}288	144 ⯃	24
triprism			192 △	96
hexagon		(288) ∥	(288) □	96
Truncated Hypercuboctahedral				
tr octahedron	192	^{2}288	192 ⬡	24
cube		^{3}96	144 □	24
3-lines			(576) ∥	96
hexagon			(288) ∥	96
Truncated Hyperoctahedral				
tr octahedron	576	1152	192 △	24
rhombicuboctahedron		^{2}288	192 ⬡	24
triprism			288 □	96
hexaprism			144 □	96
		(288) ∥	(288) □	
Hyperhombicuboctaprism				

Table 31: More Dodecadodecaskews

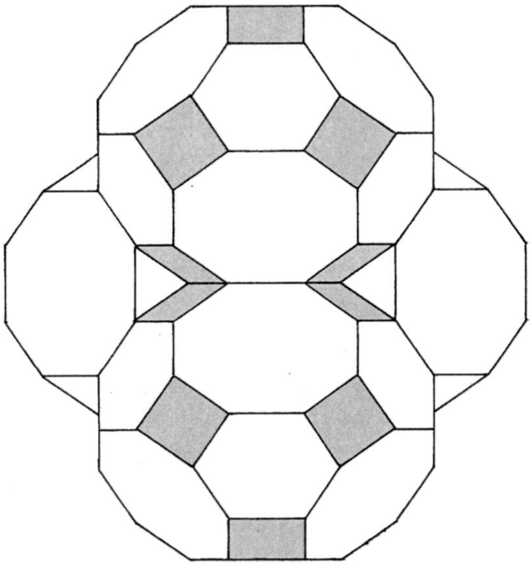

[tt^2{3,4,3}]

Truncated Hypercuboctahedral Dodecadodecaskew

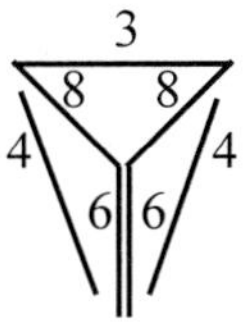

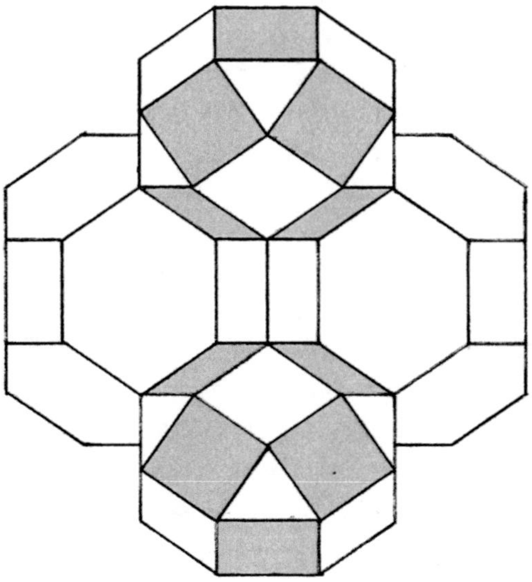

[o^2{3,4,3}]

Hyperhombi-cuboctaprism Dodecadodecaskew

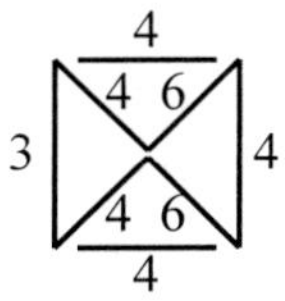

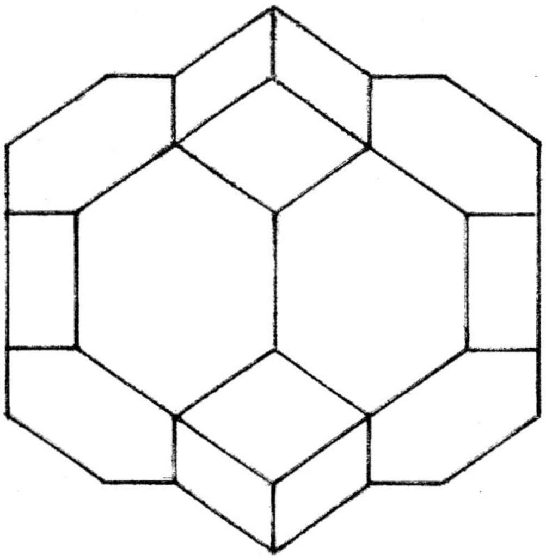

[t{3,4,3}]

Truncated Hyperoctahedral Dodecadodecaskew

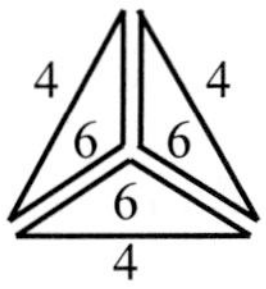

Fig.47: More Dodecadodecaskews

THE HYPERCUBE TORUS

Taking the hypercube's tetrahedral vertex figure as a potential central member, we can generate a full grid by implementing the usual vertex transformations used thus far, as shown in Figure 48 adjoining. Because of the inherent symmetry in the starting central vertex figure, there are only two further forms, one intermediate and one resultant.

The central member itself is based around the hypercube and can be envisaged in three dimensions as a torus, the doughnut shaped surface of topology, its surface entirely tiled with squares. Each of the eight cubes making up the hypercube has in effect two opposing square faces missing, forming a square tube.

Now a cube can be illustrated in two dimensions as a small square within a larger one with the two sets of vertices connected radially by further edges that form four distorted squares, a kind of perspective view. In the same way, a hypercube can be seen as a small cube within a larger one, again with the vertices connected radially, this time forming six truncated pyramidal distorted cubes between the original two.

The torus surface of this particular central hyperskew can most easily be seen as the surface comprising sixteen squares around four of the distorted pyramid cubes as shown at the top of Figure 49 overleaf.

This appears here in three dimensions to enclose an 'internal' space, but what is not immediately apparent is that the other space comprising the central cube plus two distorted pyramidal cubes funnelling into it and the external cube together comprise an exactly equivalent 'external' space.

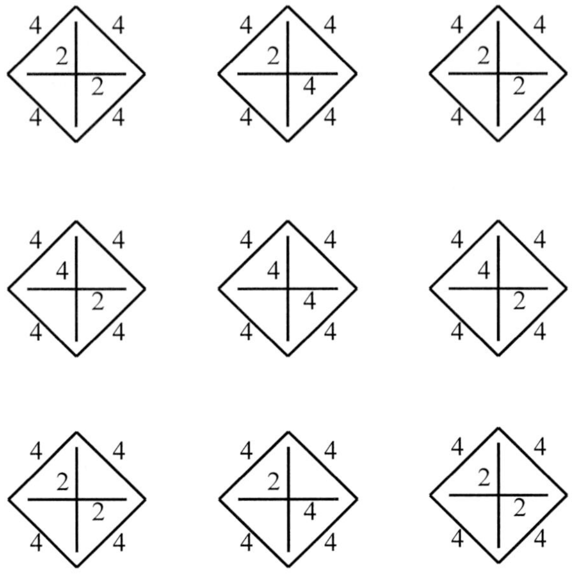

Fig.48: Didiskew Grid [2-2,2-2]

The circular loop of connections 'inside' the torus is in four dimensions identical to the outside loop that goes through the central hole we see in three dimensions. The torus can thus be seen to possess in four dimensions far greater symmetry, than is apparent in three.

These two connection loops are preserved in both the intermediate and resultant forms, although partly reduced to the form of sleeves between pairs of back to back squares. These are shown opposite at the bottom of Figure 49, with their numerical properties described in Table 32.

Because they are comprised of tubes of one kind or another, these simple hyperskews can be considered as [2-2,2-2] connected and thus named as 'Didiskews'. The Euler Number here (V-E+F) is 0, which is entirely correct for any tiling of a torus.

DIDISKEWS

Solids	V	E	F	S
cube	16	32	16 □	4
cube			(8) □	4
{4,4 |4 4}				
Hypercubic				
cube	8	16	8 □	2
square			(4) ‖	4
{4,4 |4 2}				
Cubic				
square	4	8	4 □	2
square			(4) ‖	2
{4,4 |2 2}				
Square				

Table 32: The Didiskews

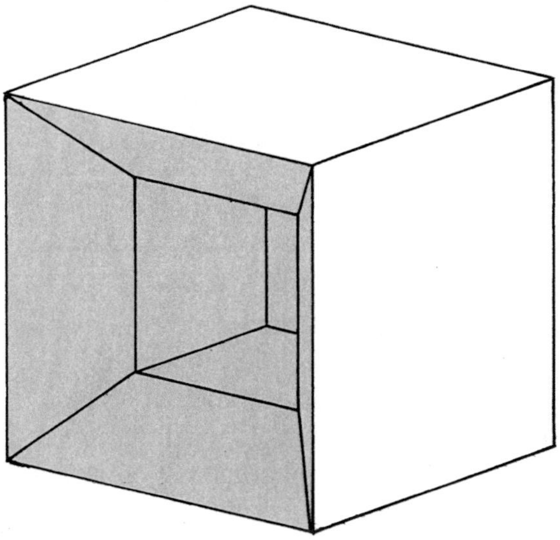

{4,4 |4 4} [{4,3,3}]

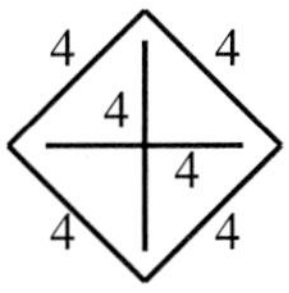

Hypercubic
Didiskew

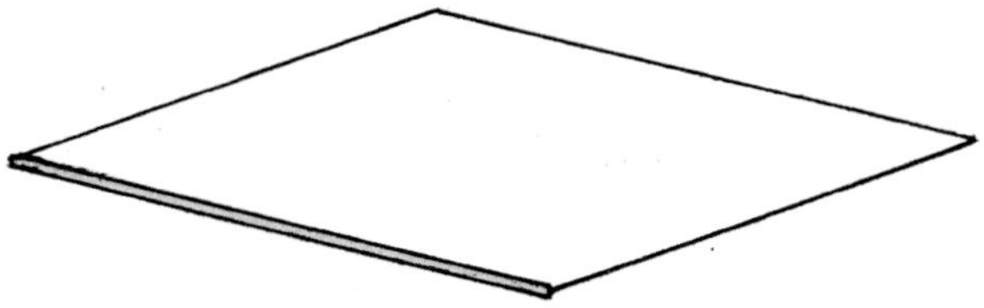

{4,4 |2 2} [{4,2,2}]

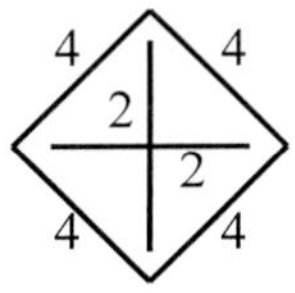

Square
Didikew

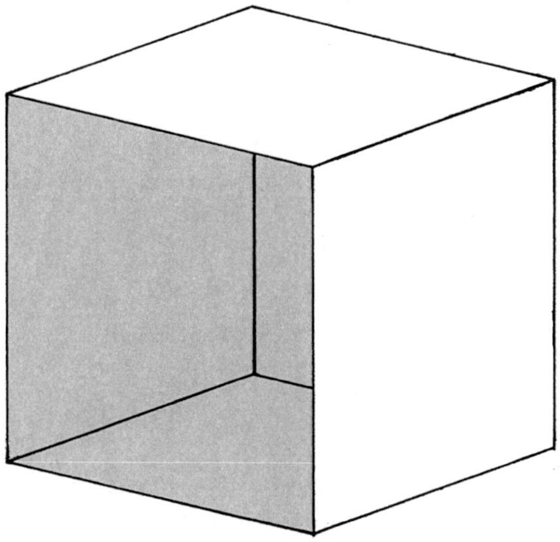

{4,4 |4 2} [{4,3,2}]

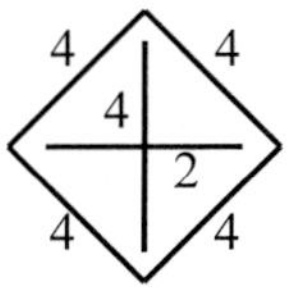

Cubic
Didiskew

Fig.49: The Didiskews

TETRAHEDRAL VERTEX FIGURES

We have now explored all of the hyperskew polyhedra that can be derived by vertex transformations from a central member with a tetrahedral vertex figure based on the various 'first odd' members of each hypersolid family, o{5,3,3}=o{3,3,5}, o{4,3,3}=o{3,3,4}, o{3,3,3} and o{3,4,3}.

Each of these has been presented in up to three possible orientations, omitting different paired faces and they have all produced family grids of hyperskews as surfaces that partition 4-dimensional space into two, each with a characteristic pattern of connectivity.

Within some of these grids there were intermediate hyperskews with tetrahedral vertex figures, each based on the first truncation of a second series (tt^2) of the hypersolid family concerned.

Except in the case of tt^2{3,3,4}, these all had an odd valued face within each vertex, which prevented vertex transformation, leaving us with an incomplete grid, essentially just a repeated single row or column of the original family grid.

The special case of tt^2{3,3,4} gave rise to a full grid (p.24) with a similar repeated column plus three new hexatetraskews as shown in Figure 21, themselves repeated within the new grid as a column each side.

The four third truncation members (t^3) of each hypersolid family also have tetrahedral vertex figures which appeared as resultants in corner positions of the family grids mentioned above with intermediate (tt^2) members. These generally have one or two odd valued faces at each vertex and thus again fail to make further grids by vertex transformation.

Now, in a similar way that a twice truncated tetrahedron t^2{3,3} is the same as an octahedron {3,4}, the twice truncated hypersolid t^2{3,3,4} is the same as the regular hypersolid {3,4,3}, so that tt^2{3,3,4} can also be regarded as t{3,4,3}.

There are three further first truncations of the regular hypersolids with tetrahedral vertex figures, t{5,3,3}, t{4,3,3} and t{3,3,3}, but these all have three odd valued faces within each vertex and therefore fail miserably to make further grids by vertex transformation.

Finally all we have left are the original regular hypersolids, three of which do indeed have tetrahedral vertex figures. Two of these, {5,3,3} and {3,3,3}, have all six faces with odd values, which can be taken no further. However as we have just seen the Hypercube {4,3,3} has all six faces as squares, which made it worthy of further investigation by vertex transformation and yielded the Didiskews (pages 52-53).

DEGENERATE HYPERSKEWS

When dealing with the skew polyhedra in the last volume of this series, we first derived the degenerate skews by considering notional truncation applied to the truncated cubic octaskew, {8,8 |3 2} in Coxeter's notation as expanded here.

A notional second truncation of this had two plus two octagons at each vertex giving us {8,4 |4 2}, a parent degenerate skew whose family of related forms are all based on the square tiling, considered as {4,4,2}, and its truncations, some of which involve prisms.

{6,4 |6 2} and {12,4 |3 2} followed as parents to families based on the hexagon and triangle tilings respectively, {6,3,2} and {3,6,2}, which also included prismatic forms that gave the two dimensional tilings a limited third dimensional presence.

The equivalent degenerate skew figures for the hypersolids are based around the regular polyhedra, seen as four dimensional figures, and their truncations. These either have zero thickness in a fourth dimension, but also include prismatic forms with a limited thickness in this dimension.

They thus share the symmetries of the dodecahedron {5,3,2}, icosahedron {3,5,2}, cube {4,3,2}, octahedron {3,4,2} and tetrahedron {3,3,2}, regarded as four dimensional forms.

Vertex figures for all the parent degenerate hyperskews are shown adjoining in Figure 50, the left hand column with a prismatic form based on one of the platonic solids, the duals of these in the right hand column based on first truncations of the dual platonic solids. We will now examine each family in turn, finding within each common connectivities between the respective solids involved.

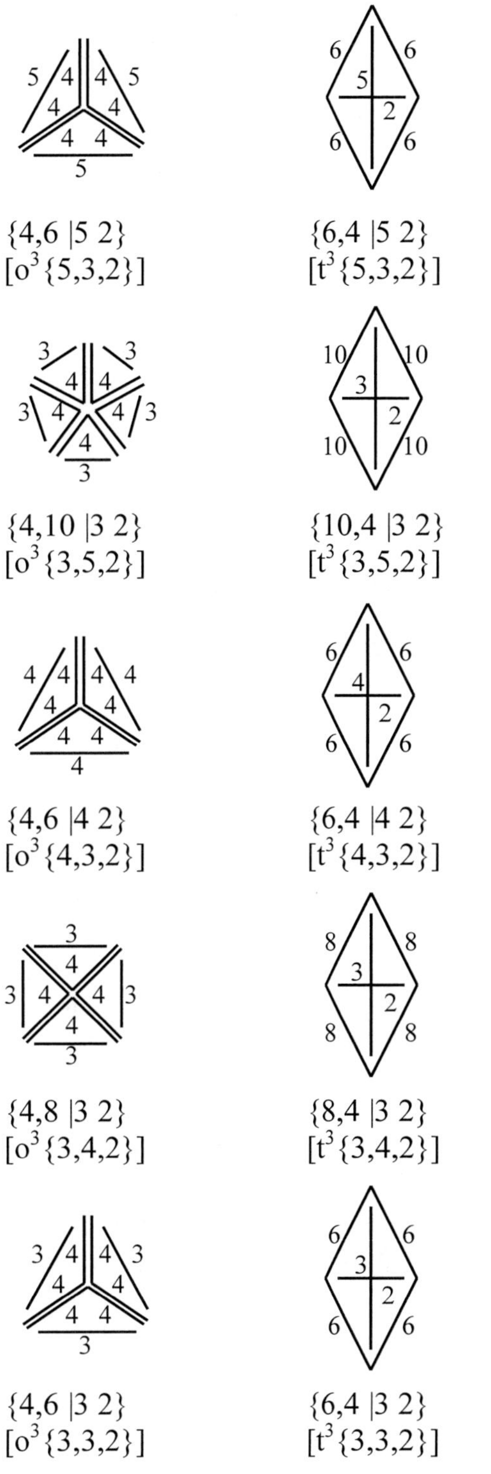

Fig.50: Parent degenerate hyperskews

DODECATRISKEW GRID

Starting with the dodecahedron considered as a four dimensional polytope, {5,3,2}, it can be thought of as having dodecahedral cells {5,3}, two per edge. These two dodecahedra back to back are thus a higher dimensional version of the dihedra we examined following the polyhedra in the appendix to *'The Simpler? Polyhedra'*. In four dimensions, viewed from three, there appear to be two dodecahedra, an inner and an outer one that share all their faces, a direct analogy to the dihedra that are back to back polygons.

The grid for this family, shown adjoining in Figure 51, has a central member based on the first odd, o{5,3,2} and two parents top right and bottom left based on $t^3\{5,3,2\}$ and $o^3\{5,3,2\}$] respectively. These three are shown opposite in Figure 52, and their numerical properties are set out in Table 33.

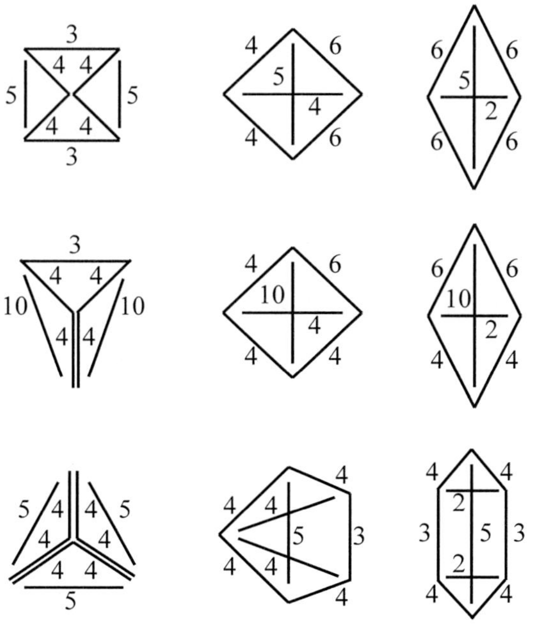

Fig.51: Dodecatriskew Grid [2-3,2-12]

The truncated icosahedral dodecatriskew {6,4 |5 2}, based on $t^3\{5,3,2\}$, comprises just 40 hexagons in back to back pairs, four per vertex and 3-connected to each other around the sphere via missing digons. The two truncated icosahedra, one inner one outer, are 12-connected to each other via the missing pentagon faces.

Its dual, the dodecaprism dodecatriskew {4,6 |5 2}, based on $o^3\{5,3,2\}$, exhibits the same connectivities. Each 'three lines' is 3-connected to paired back to back squares (digon prisms in the form of hollow sleeves) linking around the sphere, whilst the hollow pentagon prisms between the inner and outer dodecahedra provide the necessary 12-connections.

The truncated icosidodecaprism dodecatriskew, based on o{5,3,2}, central to this grid, is similarly connected as can more readily be seen in the illustration with its hexaprisms 3-connected to each other by cubes and its 12-connected truncated icosidodecahedra linked by decaprisms.

As before all members of this family grid share the 3- and 12-connectivity, along with a common Euler Number of -20.

The remainder of this family are shown overleaf. Figure 53 on page 58 shows the two intermediates and a resultant from the bottom right of the grid, these forms being detailed in the adjoining Table 34, whilst Figure 54 on page 59 shows those at top left, detailed in Table 35.

DODECATRISKEWS

Solids	V	E	F	S
hexagon	60	120	40 ⬡	20
tr icosahedron			(12) ⬠	2
{6,4 |5 2}			(30) ‖	
Truncated Icosahedral				
hexaprism	240	480	40 ⬡	20
tr icosidodecahedron			180 □	2
decaprism			(24) ⬡	12
cube			(30) □	30
nt^2t^2{4,6 |5 2}				
Truncated Icosidodecaprism				
3-lines	40	120	60 □	20
dodecahedron			(24) ⬠	2
pentaprism			(30) ‖	12
square				30
{4,6 |5 2}				
Dodecaprism				

Table 33: Dodecatriskews

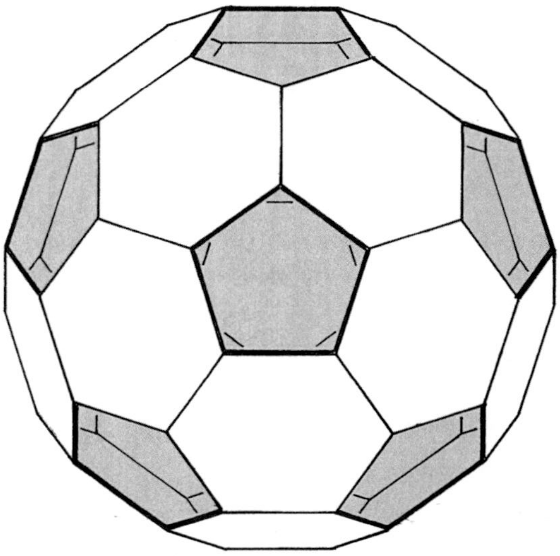

{6,4 |5 2} [t^3{5,3,2}] [t{3,5,2}]
Truncated Icosahedral Dodecatriskew

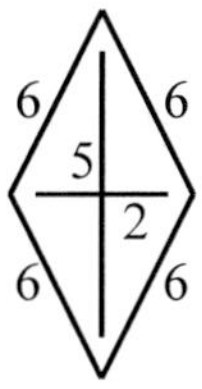

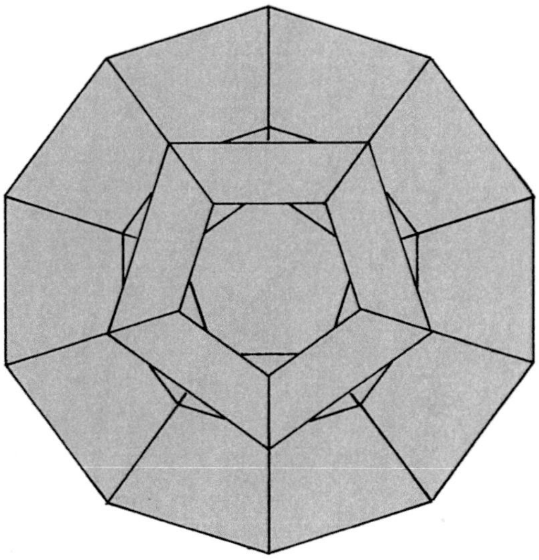

{4,6 |5 2} [o^3{5,3,2}] [t{2,5,3}]
Dodecaprism Dodecatriskew

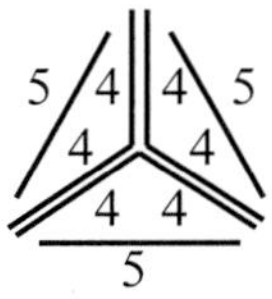

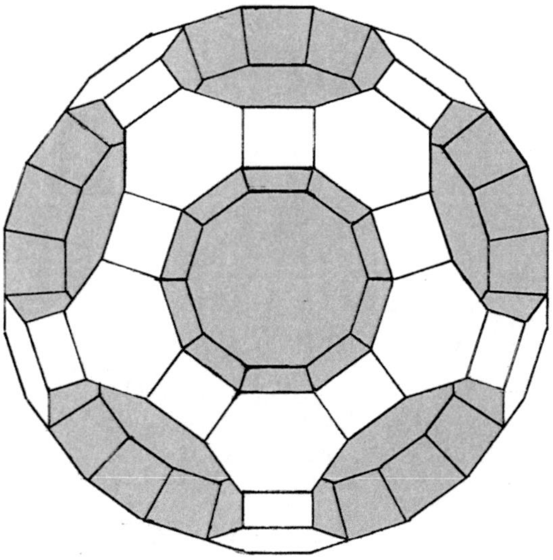

nt^2t^2{4,6 |5 2} [o{5,3,2}]
Truncated Icosidodecaprism Dodecatriskew

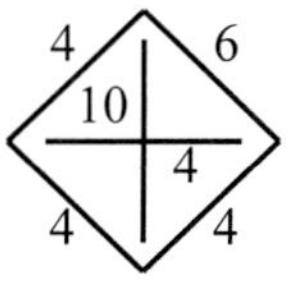

Fig.52: Dodecatriskews

MORE DODECATRISKEWS

Solids	V	E	F	S
hexagon	120	240	40 ⬡	20
tr icosidodeca			60 □	2
square			(12) ◯	30
			(60) ‖	
Truncated Icosidodecahedral				
triangle	60	180	40 △	20
rhombicosidodeca			60 □	2
square			(12) ⬠	30
			(60) ‖	
Rhombicosidodecahedral				
triprism	120	300	40 △	20
rhombicosidodeca			120 □	2
pentaprism			(24) ⬠	12
cube			(60) □	30
Rhombicosidodecaprism				

Table 34: More Dodecatriskews

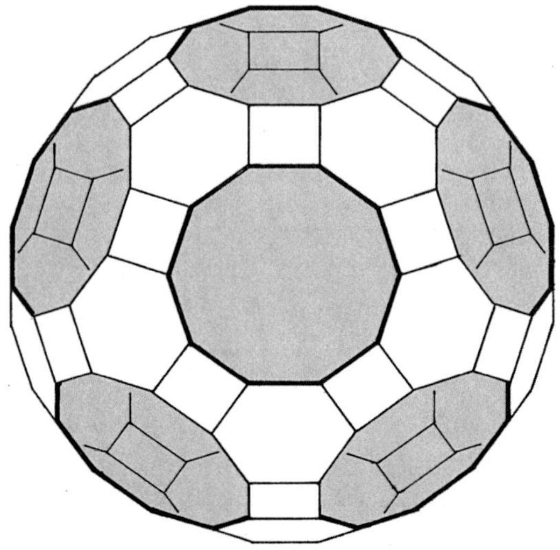

[tt^2{5,3,2}]

Truncated
Icosidodecahedral
Dodecatriskew

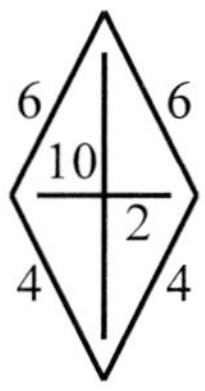

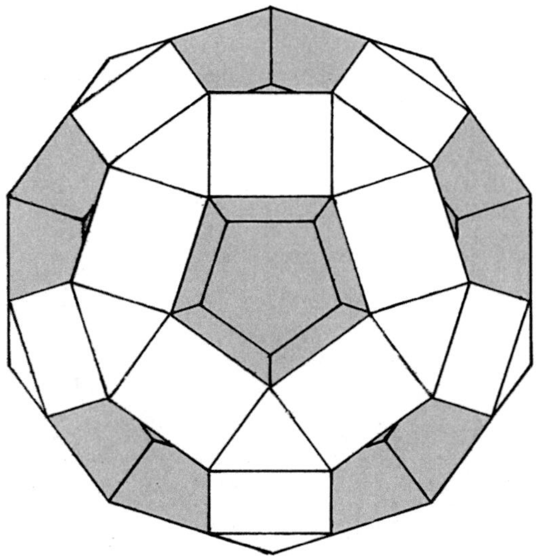

[o^2{5,3,2}]

Rhombicosidodecaprism
Dodecatriskew

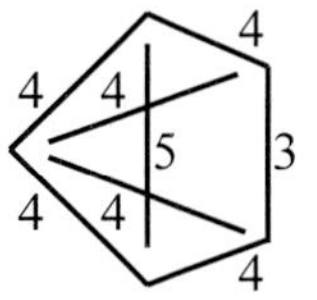

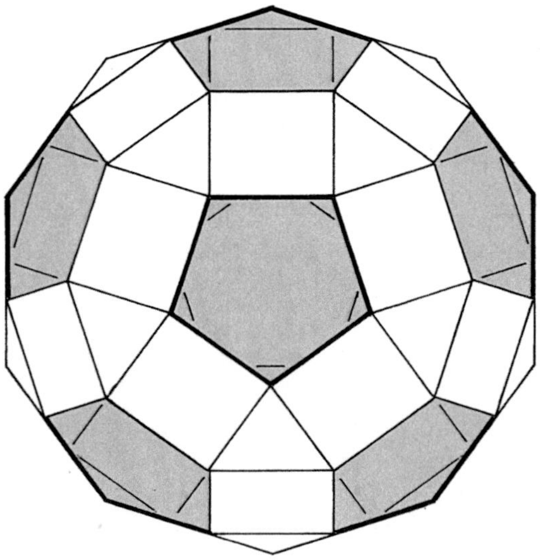

[t^{2}t^2{5,3,2}]

Rhombicosidodecahedral
Dodecatriskew

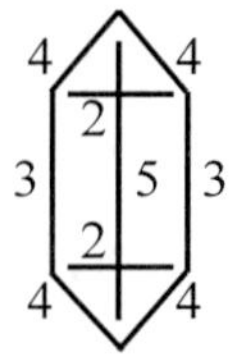

Fig.53: More Dodecatriskews

EVEN MORE DODECATRISKEWS

Solids	V	E	F	S
hexaprism	120	240	40 ⬡	20
tr icosahedron			60 □	2
pentaprism			(24) ⬠	12
			(30) □	
Truncated Icosaprism				
triprism	60	180	40 △	20
icosidodecahedron			60 □	2
pentaprism			(24) ⬠	12
			(30) ∥	
Icosidodecaprism				
triprism	120	300	40 △	20
tr dodecahedron			120 □	2
decaprism			(24) ○	12
square			(60) ∥	30
Truncated Dodecaprism				

Table 35: Even More Dodecatriskews

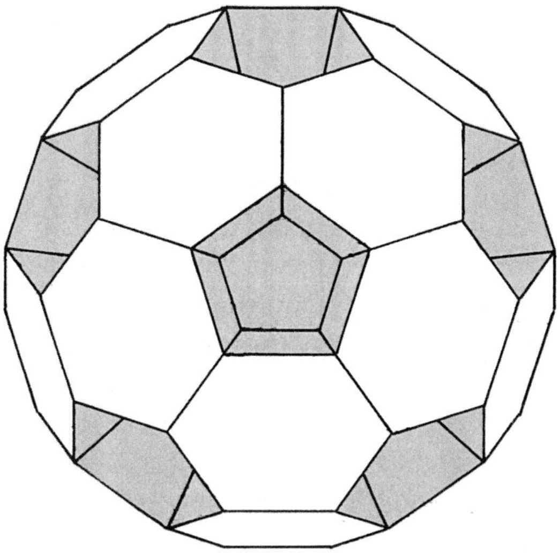

$[tt^2\{2,3,5\}]$

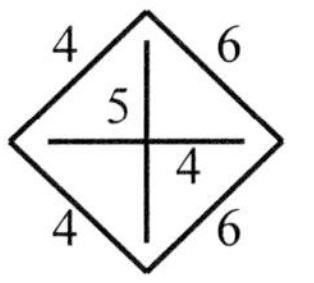

Truncated Icosaprism Dodecatriskew

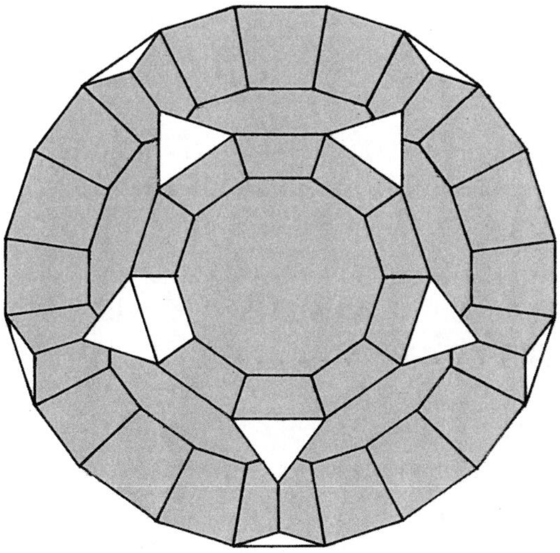

$[o^2\{2,3,5\}]$

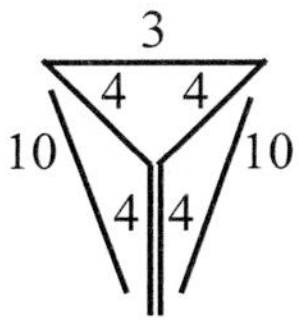

Truncated Dodecaprism Dodecatriskew

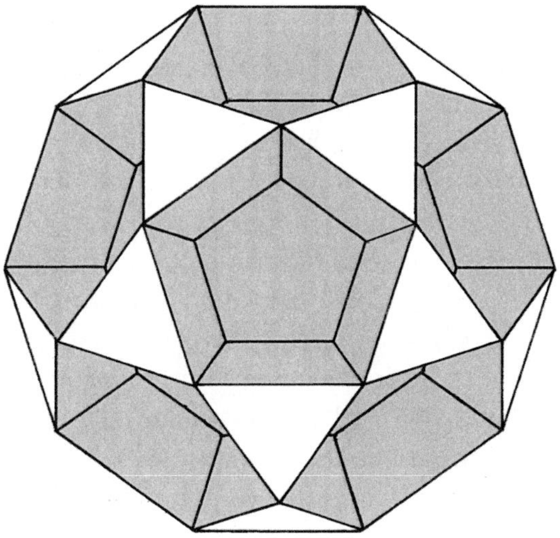

$[t^2t^2\{2,3,5\}]$

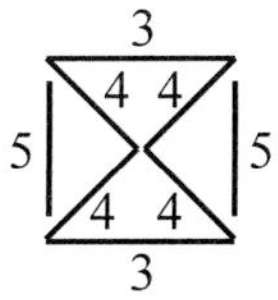

Icosidodecaprism Dodecatriskew

Fig.54: Even More Dodecatriskews

ICOSAPENTASKEW GRID

Turning now to the icosahedron considered as a four-dimensional polytope, {3,5,2} will have two icosahedral cells back to back, two per edge.

We can generate another family of degenerate hyperskews by creating a grid based on its first odd, o{3,5,2}, which with its missing hexagons and squares is simply a realignment of the central member of the previous family based on the dodecahedron.

The grid for this family, shown adjoining in Figure 55, with a central member based on that first odd, has two parents top left and bottom right based on t^3 and o^3{3,5,2} respectively. These three are shown opposite in Figure 56, and their numerical properties are set out in Table 36.

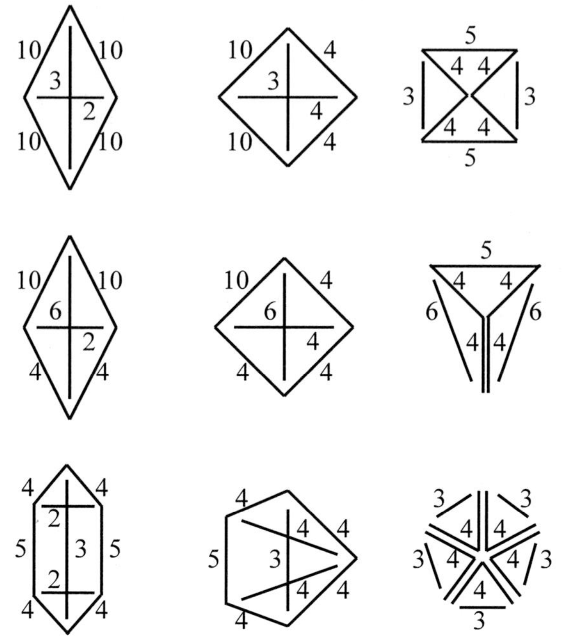

Fig.55: Icosapentaskew Grid [2-5,2-20]

The truncated dodecahedral icosapentaskew {10,4 |3 2} comprises only 24 decagons in back to back pairs, four per vertex and 5-connected to each other around the sphere via missing digons. The two truncated dodecahedra, one inner one outer, are 20-connected to each other via triangle faces.

Its dual, the icosaprism icosapentaskew {4,10 |3 2} exhibits the same connectivities. Each 'five lines' is 5-connected to paired back to back squares (digon prisms, again in the form of hollow sleeves) linking around the sphere, whilst the hollow triangle prisms between the inner and outer icosahedra provide the 20-connections.

The truncated icosidodecaprism icosapenta-skew, central to this grid, is similarly connected as can more readily be seen in the illustration with its decaprisms 5-connected to each other by cubes and its 20-connected truncated icosidodecahedra linked by hexaprisms.

The degenerate hyperskews produced by this grid are all 20- and 5-connected, so warrant the repeated use of the term 'Icosapentaskews'. As degenerates though they are far less complex, with a Euler Number of only -36.

Figure 57 and Table 37 on page 62 show the bottom left members of this grid, two intermediates based on the tt^2 and o^2{3,5,2} and the resultant in the corner based on t^2t^2{3,5,2}.

Similarly Figure 58 and Table 38 on page 63 show the top right members of the same grid, two intermediates based on tt^2 and o^2 {2,5,3} and a resultant based on t^2t^2{2,5,3}.

ICOSAPENTASKEWS

Solids	V	E	F	S
decagon	60	120	24 ⬡	12
tr dodecahedron			(20) △	2
{10,4 |3 2}			(30) ‖	
Truncated Dodecahedral				
decaprism	240	480	24 ⬡	12
tr icosidodecahedron			180 □	2
hexaprism			(40) ⬡	20
cube			(60) □	30
nt^2t^2{4,10 |3 2}				
Truncated Icosidodecaprism				
5-lines	24	120	60 □	12
icosahedron			(40) △	2
triprism			(60) ‖	20
square				30
{4,10 |3 2}				
Icosaprism				

Table 36: Icosapentaskews

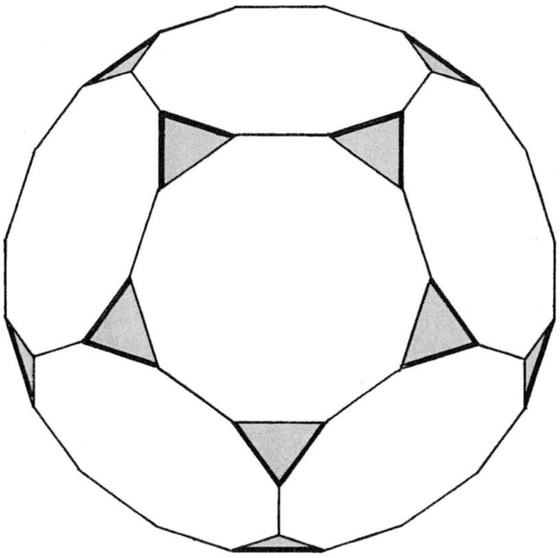

{10,4 |3 2} [t^3{3,5,2}] [t{5,3,2}]
Truncated Dodecahedral Icosapentaskew

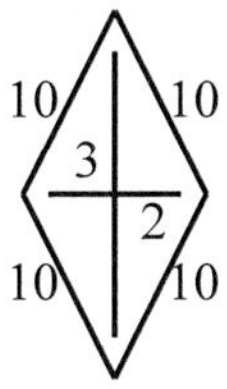

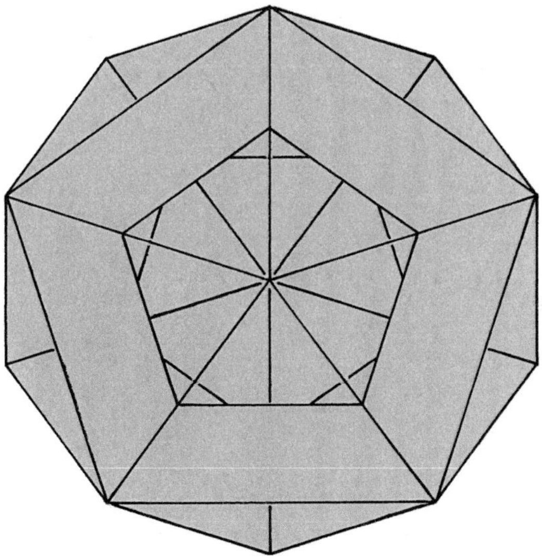

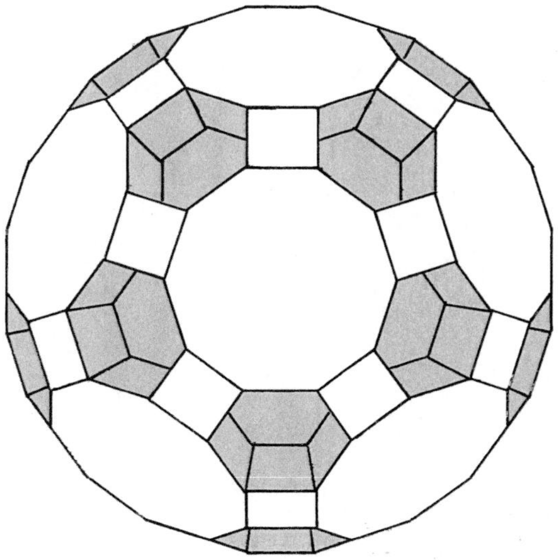

{4,10 |3 2} [o^3{3,5,2}] [t{2,3,5}]
Icosaprism Icosapentaskew

nt^2t^2{10,4 |3 2} [o{3,5,2}]
Truncated Icosidodecaprism Icosapentaskew

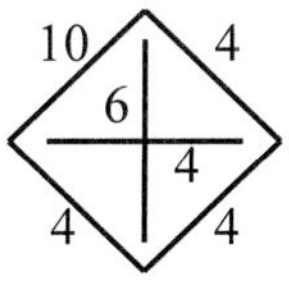

Fig.56: Icosapentaskews

MORE ICOSAPENTASKEWS

Solids	V	E	F	S
decagon	120	240	24 ⬡	12
tr icosidodeca			60 □	2
square			(20) ⬡	30
			(60) ∥	
Truncated Icosidodecahedral				
pentagon	60	180	24 ⬠	12
rhombicosidodeca			60 □	2
square			(20) △	30
			(60) ∥	
Rhombicosidodecahedral				
pentaprism	120	300	24 ⬠	12
rhombicosidodeca			120 □	2
triprism			(40) △	20
cube			(60) □	30
Rhombicosidodecaprism				

Table 37: More Icosapentaskews

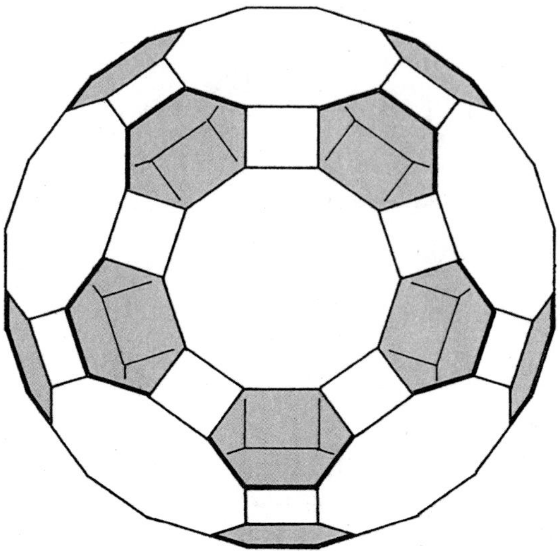

$[tt^2\{3,5,2\}]$

Truncated Icosidodecahedral Icosapentaskew

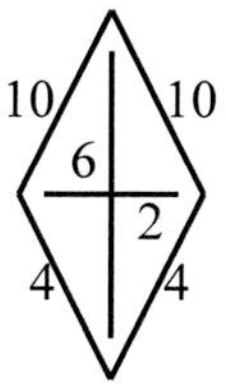

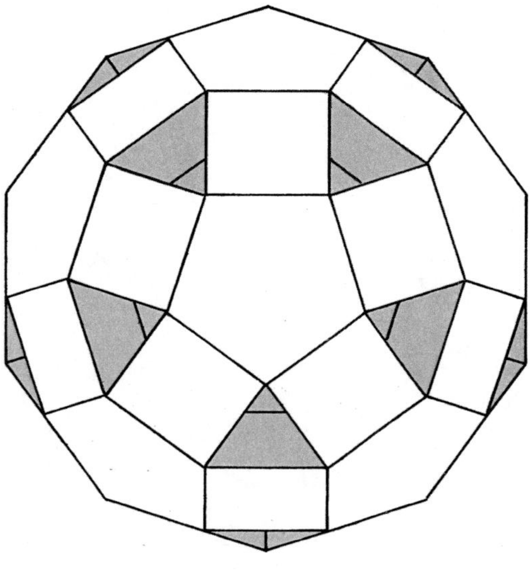

$[o^2\{3,5,2\}]$

Rhombicosidodecaprism Icosapentaskew

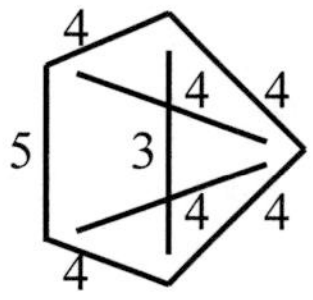

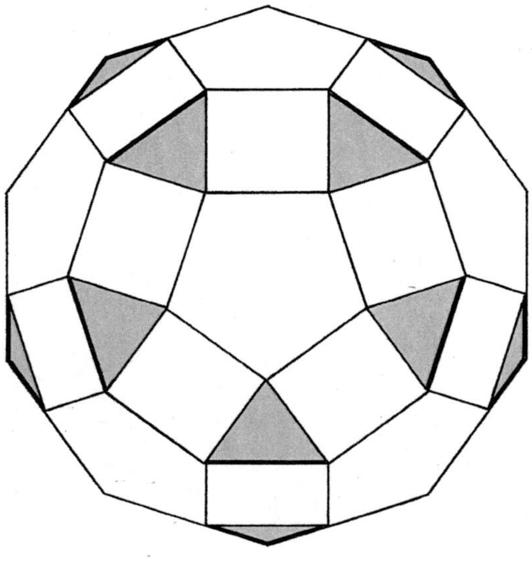

$[t^2t^2\{3,5,2\}]$

Rhombicosidodecahedral Icosapentaskew

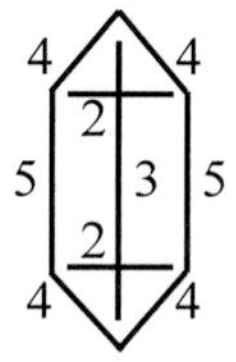

Fig.57: More Icosapentaskews

EVEN MORE ICOSAPENTASKEWS

Solids	V	E	F	S
decaprism	120	240	24 ◯	12
tr dodecahedron			60 □	2
triprism			(40) △	20
			(30) □	
Truncated Dodecaprism				
pentaprism	60	180	24 ⬠	12
icosidodecahedron			60 □	2
triprism			(40) △	20
			(30) ‖	
Icosidodecaprism				
pentaprism	120	300	24 ⬠	12
tr icosahedron			120 □	2
hexaprism			(40) ⬡	20
square			(60) ‖	30
Truncated Icosaprism				

Table 38: Even More Icosapentaskews

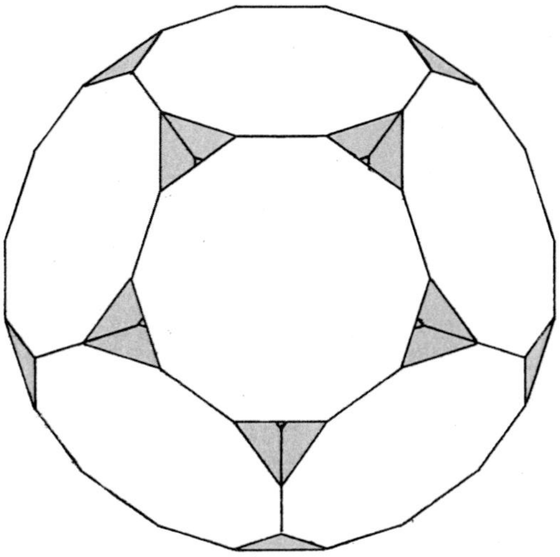

[tt²{2,5,3}]

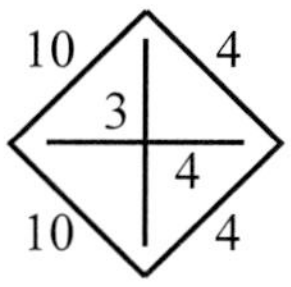

Truncated Dodecaprism Icosapentaskew

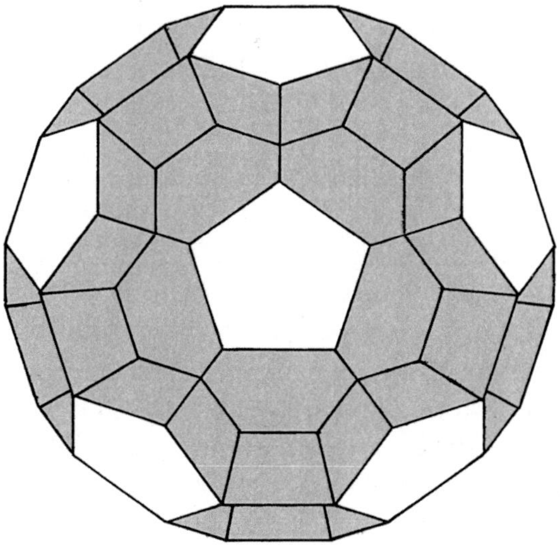

[o²{2,5,3}]

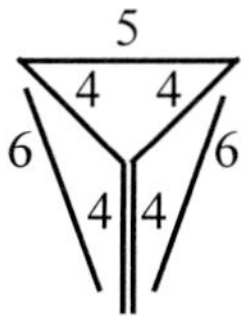

Truncated Icosaprism Icosapentaskew

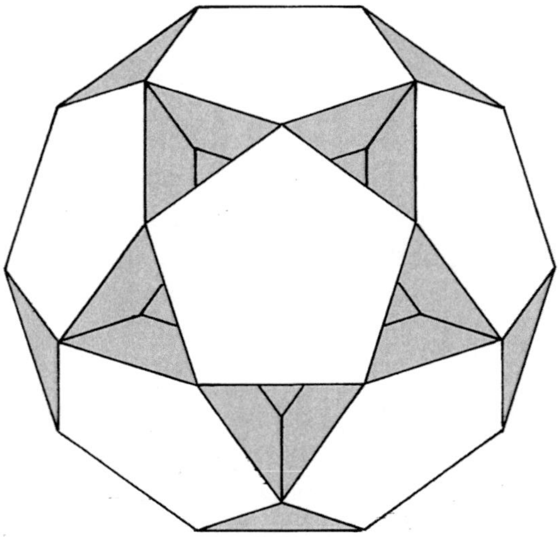

[t²t²{2,5,3}]

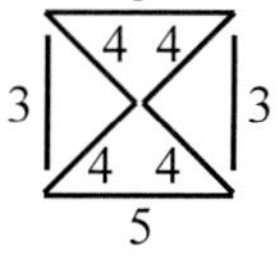

Icosidodecaprism Icosapentaskew

Fig.58: Even More Icosapentaskews

TRIACONTAPENTASKEW GRID

Finally taking the third possible alignment of the central tetrahedral vertex figure of o{5,3,2}, we can miss out the square faces of the truncated icosidodecahedra, making them 30-connected.

These link to 2-connected cubes to form one space, whilst the decagon prisms, with some missing square faces, are 5-connected to form the other space linked by 3-connected hexagon prisms.

We have accordingly called these 'Triacontapentaskews' and again this vertex can be transformed to produce another full grid of nine members as shown adjoining in Figure 59.

All of these are again topologically similar surfaces between a 30-connected space and a 5-connected one. Taking double vertices and edges into account, they also have a consistent Euler Number (V-E+F) of -56, not surprisingly the sum of the Euler Numbers of the previous two families.

Two of the resultants in this grid are also based on t^3{5,3,2} = t^3{2,3,5} and o^3{5,3,2} = o^3{2,3,5} respectively, and along with the central member are shown opposite in Figure 60, with Table 39 describing their numerical quantities.

Figure 61 and Table 40 on page 66 show the top left members of this grid, two intermediates based on the o^2{2,3,5} and tt^2{5,3,2}, and the resultant in the corner based on t{5,3,2}.

Similarly Figure 62 and Table 41 on page 67 illustrate the bottom right members of this grid, with its two intermediate forms based on tt^2{2,3,5} and o^2{5,3,2}, and the resultant in the corner based on t{2,3,5}.

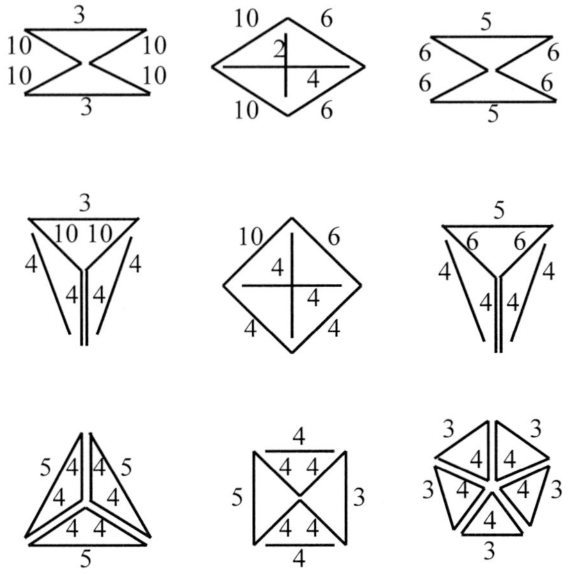

Fig.59: Triacontapentaskew Grid [3-5,2-30]

Compared to the dodecatriskews and icosapentaskews we have just seen, many more of the triacontapentaskews have sleeves of back to back squares providing the necessary connections and two are connected by slots or faceless digons.

INCOMPLETE GRIDS

We have thus far examined three grids of degenerate vertex figures covering dodecatriskews, icosapentaskews and triacontapentaskews. As with the full four dimensional hyperskews, none of the intermediate or resultant members of these grids with tetrahedral vertex figures can produce further grids (see p.16).

In general taking these vertex figures through transformation will only reproduce the row or column within which they sit. This is because the odd valued faces on their diagonals, {5} and {3}, cannot be produced by truncation and nor can the digon {2} we have in some instances here.

TRIACONTAPENTASKEWS

Solids	V	E	F	S
hexagon	60	180	24 ⬠	20
tr icosahedron			40 ⬡	2
pentagon			(60) ‖	12
digon				30
Truncated Icosahedral				
hexaprism	240	480	24 ○	20
tr icosidodecahedron			40 ⬡	2
decaprism			120 □	12
cube			(60) □	30
			(60) □	
Truncated Icosidodecaprism				
3-lines	40	180	24 ⬠	20
dodecahedron			60 □	2
pentaprism			(60) ‖	12
square			(60) ‖	30
Dodecaprism				

Table 39: Triacontapentaskews

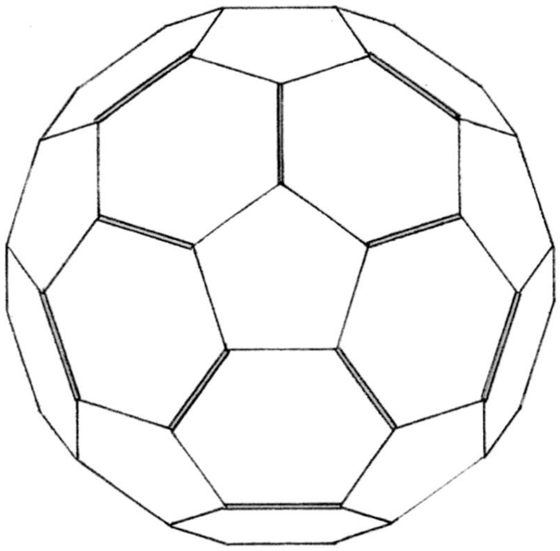

$[t^3\{5,3,2\}]$
$[t\{3,5,2\}]$

Truncated
Icosahedral
Triacontapentaskew

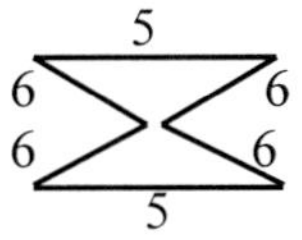

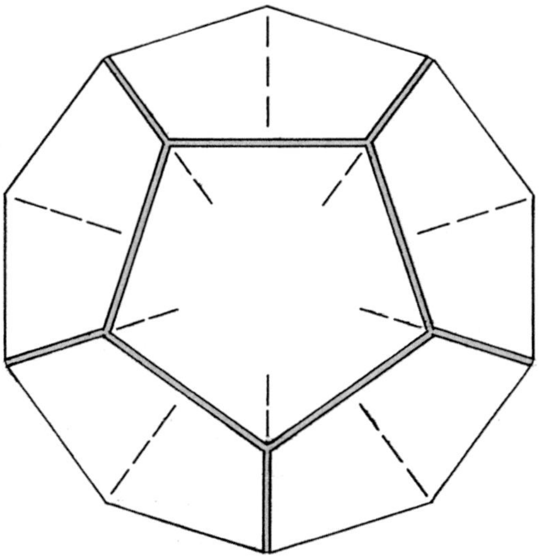

$[o^3\{5,3,2\}]$
$[t\{2,5,3\}]$

Dodecaprism
Triacontapentaskew

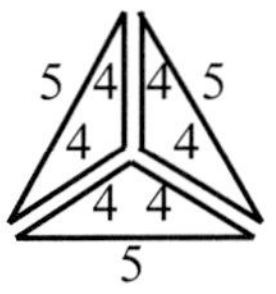

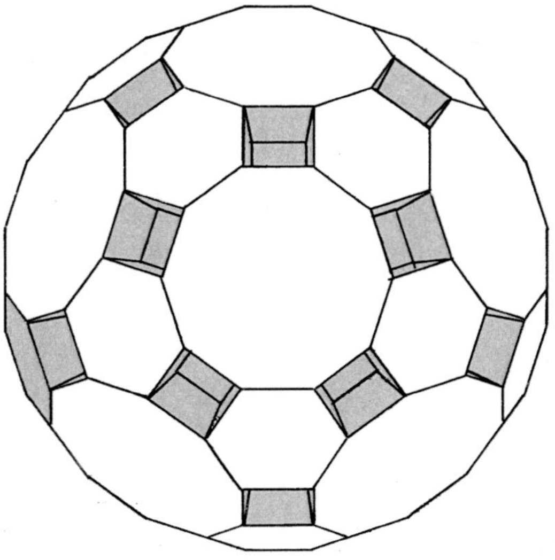

$[o\{5,3,2\}]$

Truncated
Icosidodecaprism
Triacontapentaskew

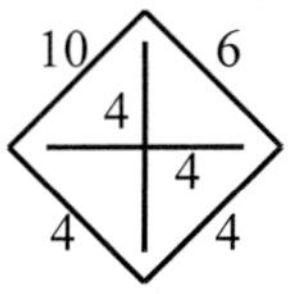

Fig.60: Triacontapentaskews

MORE TRIACONTAPENTASKEWS

Solids	V	E	F	S
hexagon	120	240	24 ◯	20
tr icosidodeca			40 ⬡	2
decagon			(30) □	12
square			(60) ‖	30
Truncated Icosidodecahedral				
triangle	60	180	24 ◯	20
tr dodecahedron			40 △	2
decagon			(60) ‖	12
digon				30
Truncated Dodecahedral				
triprism	120	360	24 ◯	20
tr dodecahedron			40 △	2
decaprism			120 □	12
square		(60) ‖	(60) □	30
Truncated Dodecaprism				

Table 40: More Triacontapentaskews

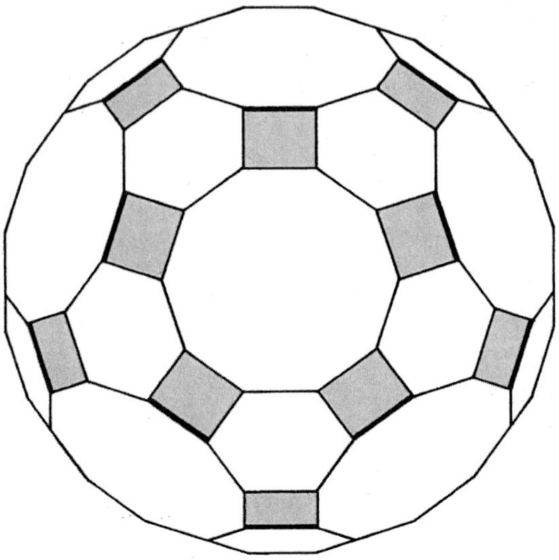

[tt²{5,3,2}]
Truncated [tt²{3,5,2}]
Icosidodecahedral
Triacontapentaskew

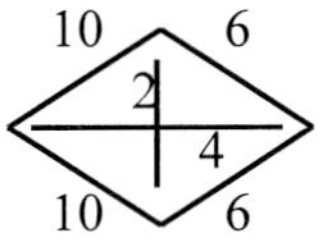

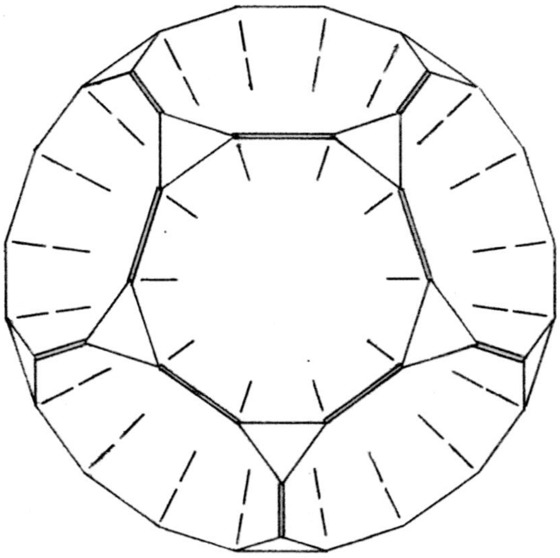

[o²{2,3,5}]
Truncated [tt²{2,5,3}]
Dodecaprism
Triacontapentaskew

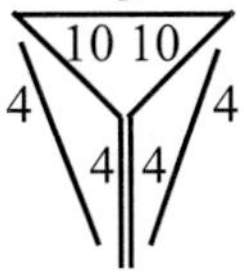

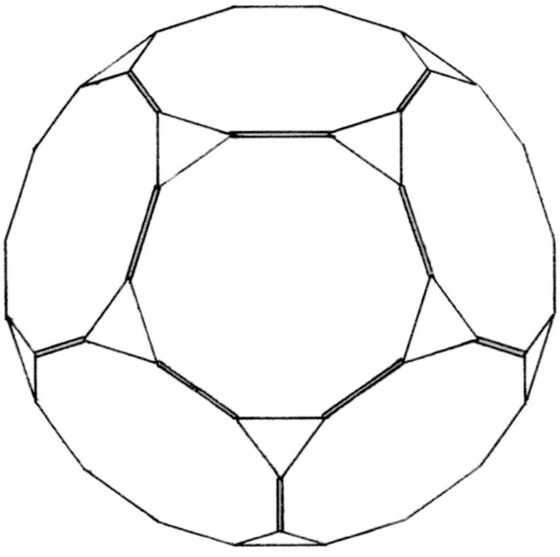

[t³{3,5,2}]
Truncated [t{5,3,2}]
Dodecahedral
Triacontapentaskew

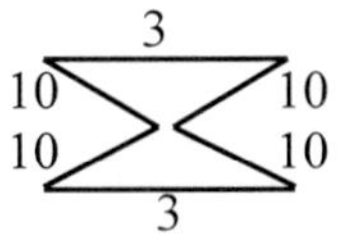

Fig.61: More Triacontapentaskews

EVEN MORE TRIACONTAPENTASKEWS

Solids	V	E	F	S
hexaprism	120	360	24 ⬠	20
tr icosahedron			40 ⬡	2
pentaprism			120 □	12
square		(60) ‖	(60) □	30
Truncated Icosaprism				
triprism	24	180	40 △	20
icosahedron			60 □	2
5-lines			(60) ‖	12
square			(60) ‖	30
Icosaprism				
triprism	120	360	24 ⬠	20
rhombicosidodeca			40 △	2
pentaprism			120 □	12
cube		(60) ‖	(60) □	30
Rhombicosidodecaprism				

Table 41: Even More Triacontapentaskews

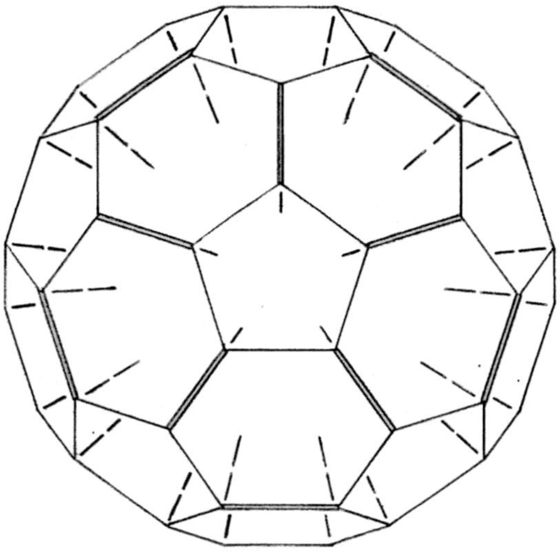

[tt²{2,3,5}]
[o²{2,5,3}]

Truncated
Icosaprism
Triacontapentaskew

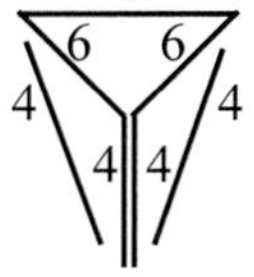

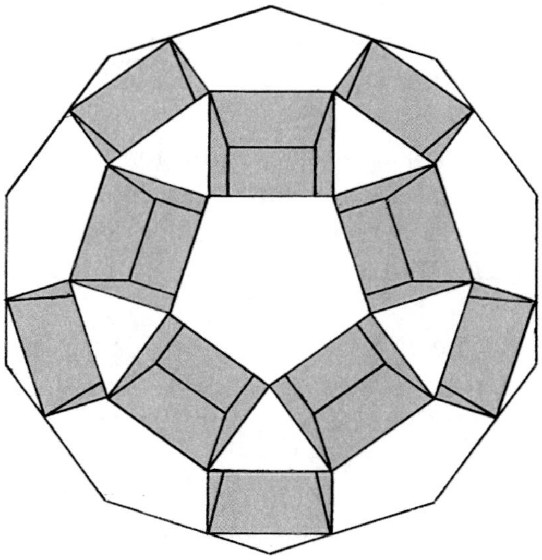

[o²{5,3,2}]
[o²{3,5,2}]

Rhombicosidodecaprism
Triacontapentaskew

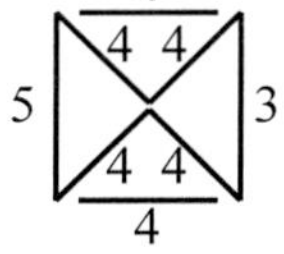

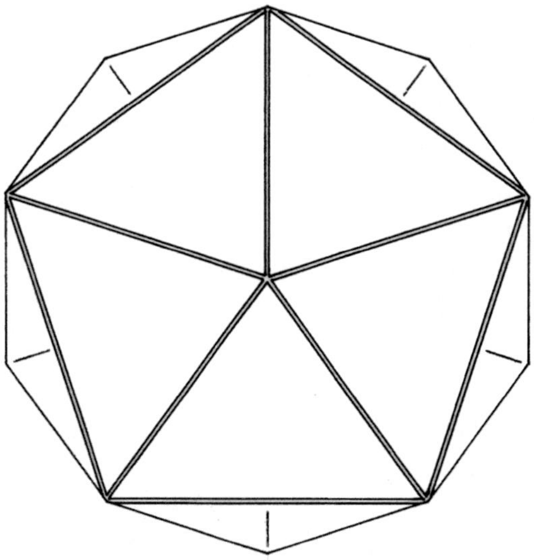

[o³{3,5,2}]
[t{2,3,5}]

Icosaprism
Triacontapentaskew

Fig.62: Even More Triacontapentaskews

DEGENERATE HEXATRISKEWS

We can now move on to consider the cube as a four dimensional polytope, {4,3,2}, comprising two cubic cells {4,3}, two per edge. These two cubes back to back are thus a higher dimensional version of the dihedra we examined following the polyhedra in the appendix to *'The Simpler? Polyhedra'*.

In four dimensions, viewed from three, there appear to be two cubes, an inner and an outer one that share all their faces, a direct analogy to the dihedra that are back to back polygons.

The grid for this family, shown adjoining in Figure 63, has a central member based on the first odd, o{4,3,2} and two parents top right and bottom left based on $t^{3}\{4,3,2\}$ and $o^{3}\{4,3,2\}$] respectively. These three are shown opposite in Figure 64, and their numerical properties are set out in Table 42.

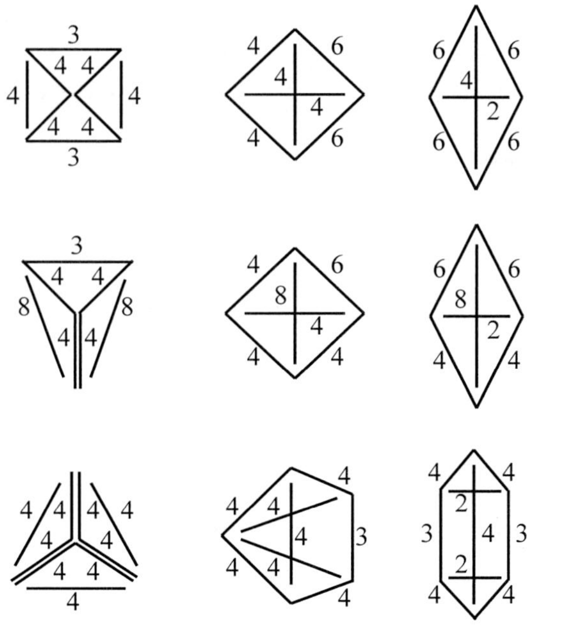

Fig.63: Hexatriskew Grid [2-3,2-6]

Shown at top right the truncated octahedral hexatriskew {6,4 |4 2} comprises only 16 hexagons in back to back pairs, four per vertex and 3-connected to each other around the sphere via missing digons. The two truncated octahedra, one inner one outer, are 6-connected to each other via the missing square faces.

Shown at bottom left, its dual, the cuboprism hexatriskew {4,6 |4 2} exhibits the same connectivities. Each 'three lines' is 3-connected to paired back to back squares (digon prisms in the form of hollow sleeves as before) linking around the sphere, whilst the hollow square prisms between the inner and outer cubes provide the necessary 6-connections.

Shown at bottom right, the truncated cuboctaprism hexatriskew, central to this grid, is similarly connected as can more readily be seen in the illustration with its hexaprisms 3-connected to each other by cubes and its 6-connected truncated cuboctahedra linked by octaprisms.

As before all members of this family grid share these same connectivities and a common Euler Number of -8.

Overleaf Figure 65 shows the two intermediates and a resultant from bottom right of the grid, as detailed in the adjoining Table 43, whilst Figure 66 facing that shows those at top left, as detailed in Table 44.

HEXATRISKEWS

Solids	V	E	F	S
hexagon	24	48	16 ⬡	8
tr octahedron			(6) □	2
{6,4 |4 2}			(12) ‖	
Truncated Octahedral				
hexaprism	96	192	16 ⬡	8
tr cuboctahedron			72 □	2
octaprism			(12) ⯃	6
cube			(24) □	12
nt^2t^2{4,6 |4 2}				
Truncated Cuboctaprism				
3-lines	16	48	24 □	8
cube			(12) □	2
cube			(24) ‖	6
square				12
{4,6 |4 2}				
Cuboprism				

Table 42: Hexatriskews

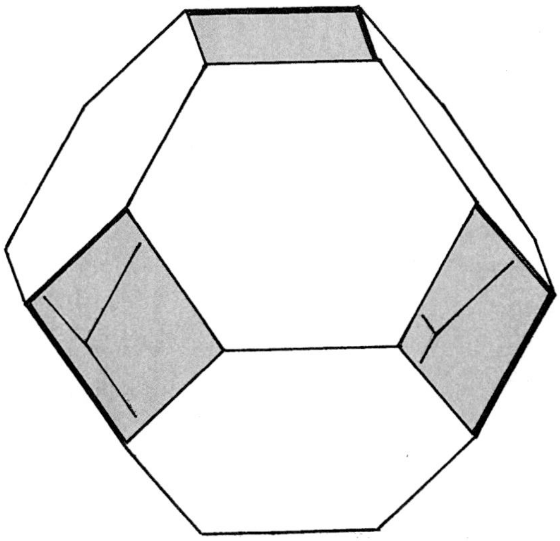

{6,4 |4 2} [t^3{4,3,2}] [t{3,4,2}]
Truncated Octahedral Hexatriskew

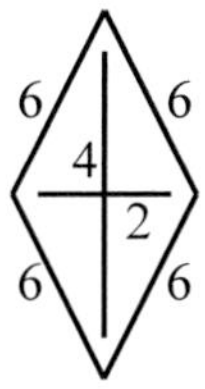

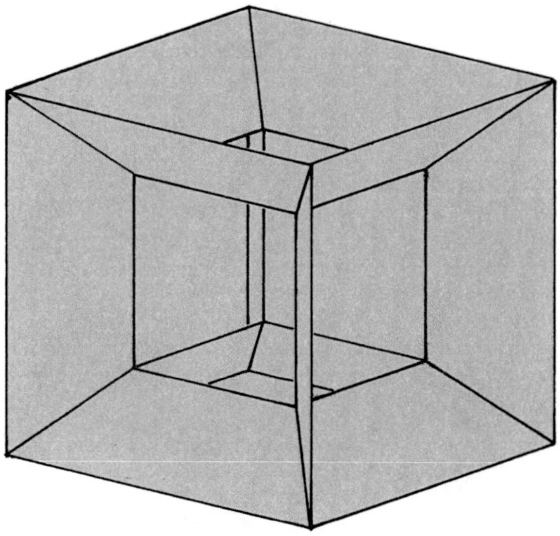

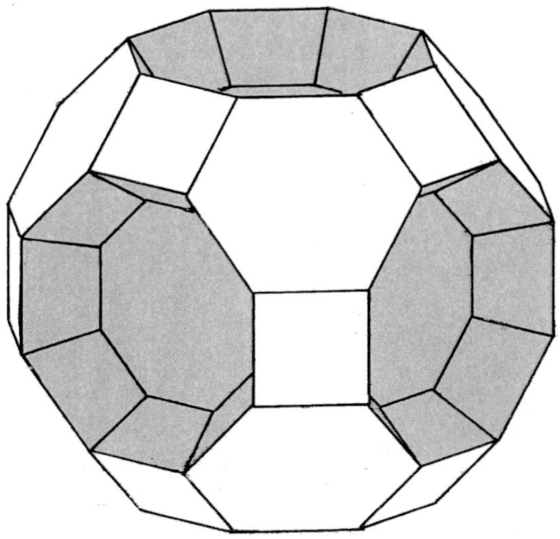

{4,6 |4 2} [o^3{4,3,2}] [t{2,4,3}]
Cuboprism Hexatriskew

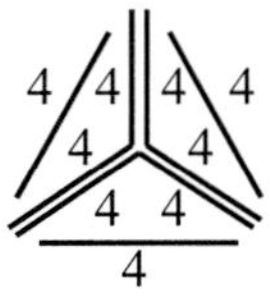

nt^2t^2{4,6 |4 2} [o{4,3,2}]
Truncated Cuboctaprism Hexatriskew

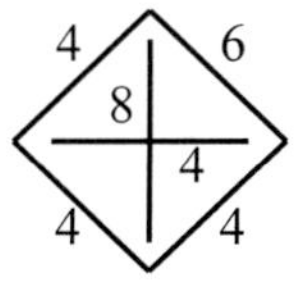

Fig.64: Hexatriskews

MORE HEXATRISKEWS

Solids	V	E	F	S
hexagon	48	96	16 ⬡	8
tr cuboctahedron			24 □	2
square			(6) ⯃	12
			(24) ∥	
Truncated Cuboctahedral				
triangle	24	72	16 △	8
rhombicuboctahedron			24 □	2
square			(6) □	12
			(24) ∥	
Rhombicuboctahedral				
triprism	48	120	16 △	8
rhombicuboctahedron			48 □	2
cube			(12) □	6
cube			(24) □	12
Rhombicuboctaprism				

Table 43: More Hexatriskews

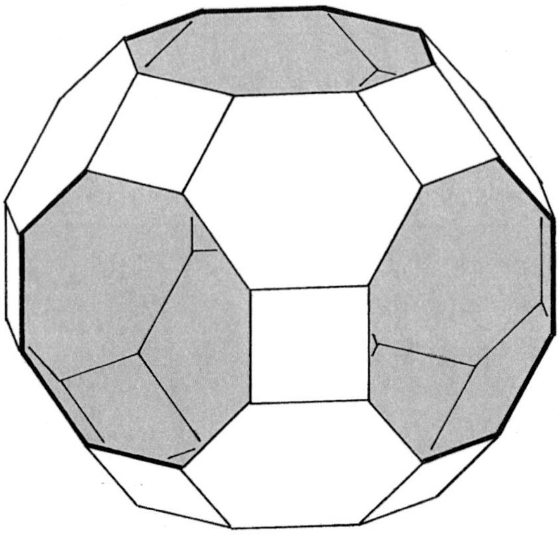

$[tt^2\{4,3,2\}]$

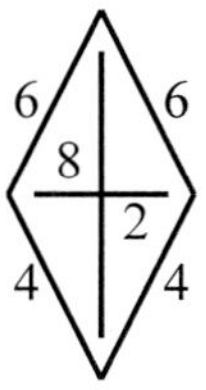

Truncated Cuboctahedral Hexatriskew

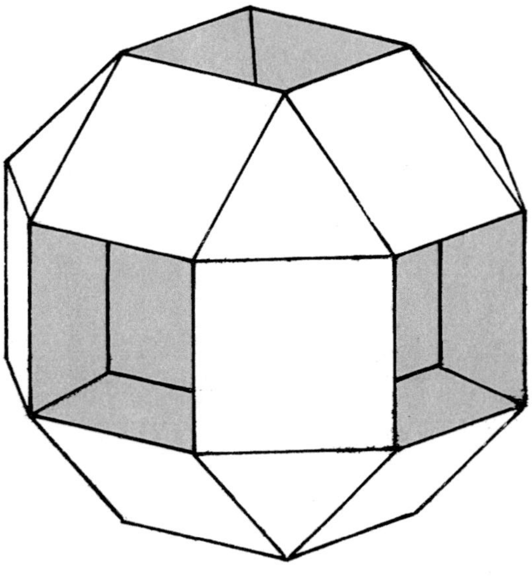

$[o^2\{4,3,2\}]$

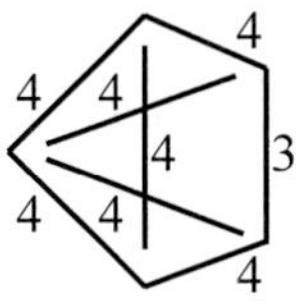

Rhombicuboctaprism Hexatriskew

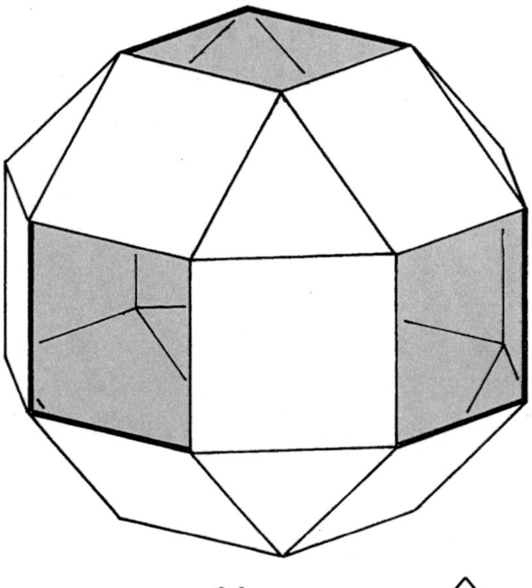

$[t^2t^2\{4,3,2\}]$

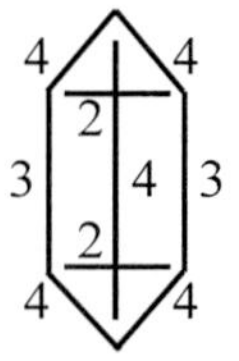

Rhombicuboctahedral Hexatriskew

Fig.65: More Hexatriskews

EVEN MORE HEXATRISKEWS

Solids	V	E	F	S
hexaprism	48	96	16 ⬡	8
tr octahedron			24 □	2
cube			(12) □	6
			(12) □	
Truncated Octaprism				
triprism	24	72	16 △	8
cuboctahedron			24 □	2
cube			(12) □	6
			(12) ‖	
Cuboctaprism				
triprism	48	120	16 △	8
truncated cube			48 □	2
octaprism			(12) ⯃	6
square			(24) ‖	12
Truncated Cuboprism				

Table 44: Even More Hexatriskews

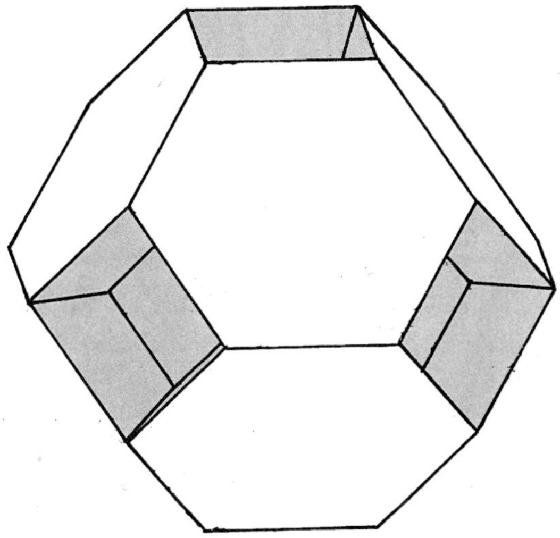

$[tt^2\{2,3,4\}]$

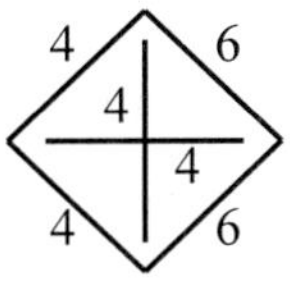

Truncated Octaprism Hexatriskew

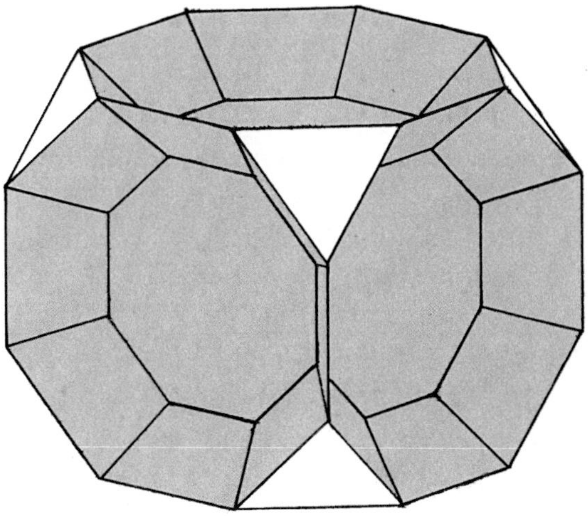

$[o^2\{2,3,4\}]$

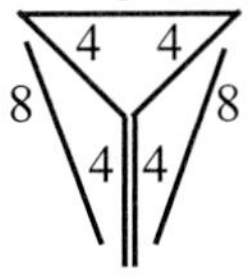

Truncated Cuboprism Hexatriskew

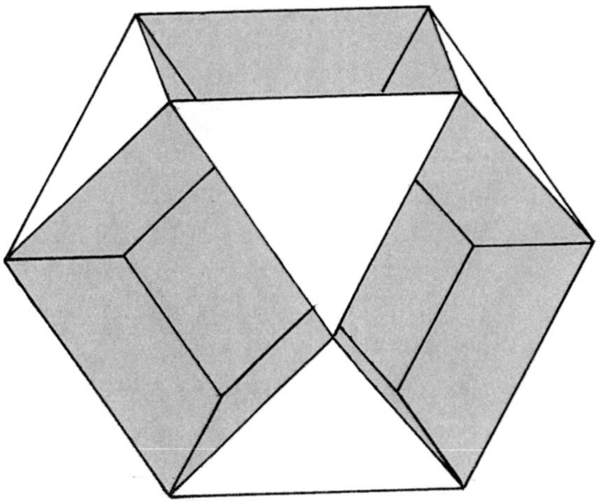

$[t^2t^2\{2,3,4\}]$

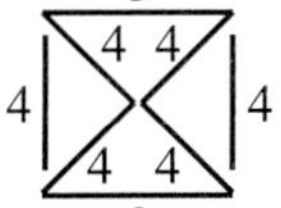

Cuboctaprism Hexatriskew

Fig.66: Even More Hexatriskews

OCTATETRASKEW GRID

Turning now to the octahedron considered as a four-dimensional polytope, {3,4,2} will have two octahedral cells back to back, two per edge.

We can generate another family of degenerate hyperskews by creating a grid based on its first odd, o{3,4,2}, which with its missing hexagons and squares is simply a realignment of the central member of the previous family based on the cube.

The grid for this family, shown adjoining in Figure 67, with a central member based on that first odd, has two parents top left and bottom right based on t^3{3,4,2} and o^3{3,4,2} respectively. These three are shown opposite in Figure 68, and their numerical properties are set out in Table 45.

The truncated cubic octatetraskew {8,4 |3 2} comprises just 12 octagons in back to back pairs, four per vertex and 4-connected to each other around the sphere via missing digons. The two truncated cubes, one inner one outer, are 8-connected to each other via missing triangle faces.

Its dual, the octaprism octatetraskew {4,8 |3 2} exhibits the same connectivities. Each 'four lines' is 4-connected to paired back to back squares (digon prisms) linking around the sphere, whilst the hollow triangle prisms between the inner and outer octahedra provide the 8-connections.

The truncated cuboctaprism octatetraskew, central to this grid, is similarly connected as can more readily be seen in the illustration with its octaprisms 4-connected to each other by cubes and its 8-connected truncated cuboctahedra linked by hollow hexaprisms.

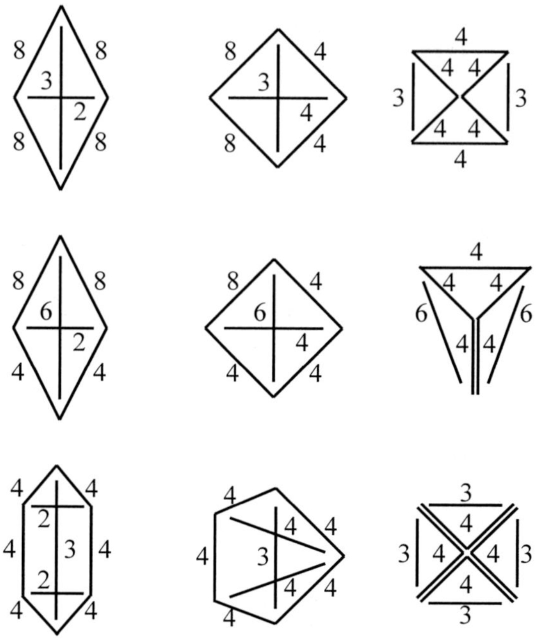

Fig.67: Octatetraskew Grid [2-4,2-8]

The degenerate hyperskews produced by this grid are all 8- and 4-connected, hence the use of the term 'octatetraskews', sharing a common Euler Number of only -12.

Figure 69 and Table 46 on page 74 show the bottom left members of this grid, two intermediates based on the tt^2 and o^2{3,4,2} and the resultant in the corner based on t^2t^2{3,4,2}.

Similarly Figure 70 and Table 47 on page 75 show the top right members of the same grid, two intermediates based on tt^2 and o^2 {2,4,3} and a resultant based on t^2t^2{2,4,3}.

OCTATETRASKEWS

Solids	V	E	F	S
octagon truncated cube {8,4 |3 2} Truncated Cubic	24	48	12 ⯃ (8) △ (12) ‖	6 2
octaprism tr cuboctahedron hexaprism cube nt^2t^2{4,8 |3 2} Truncated Cuboctaprism	96	192	12 ⯃ 72 □ (16) ⬡ (24) □	6 2 8 12
4-lines octahedron triprism square {4,8 |3 2} Octaprism	12	48	24 □ (16) △ (24) ‖	6 2 8 12

Table 45: Octatetraskews

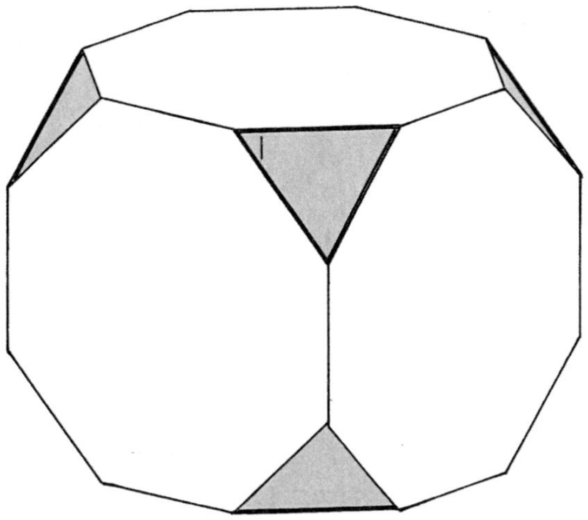

{8,4 |3 2} $[t^3\{3,4,2\}]$ [t{4,3,2}]
Truncated Cubic Octatetraskew

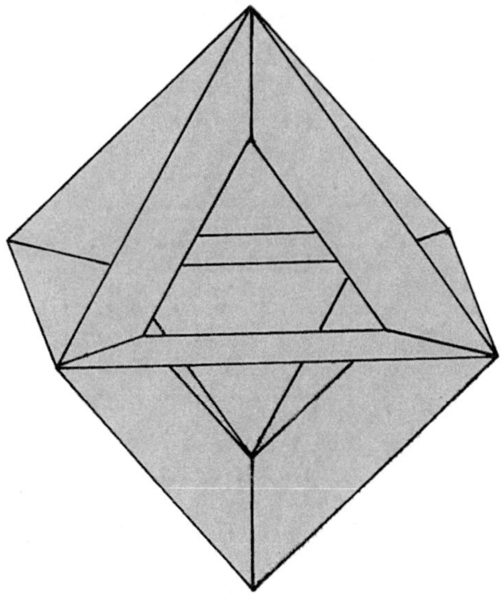

{4,8 |3 2} $[o^3\{3,4,2\}]$ [t{2,3,4}]

Octaprism Octatetraskew

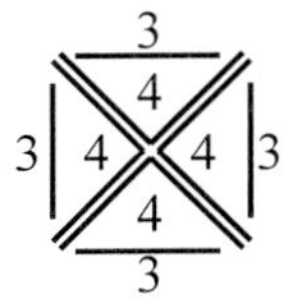

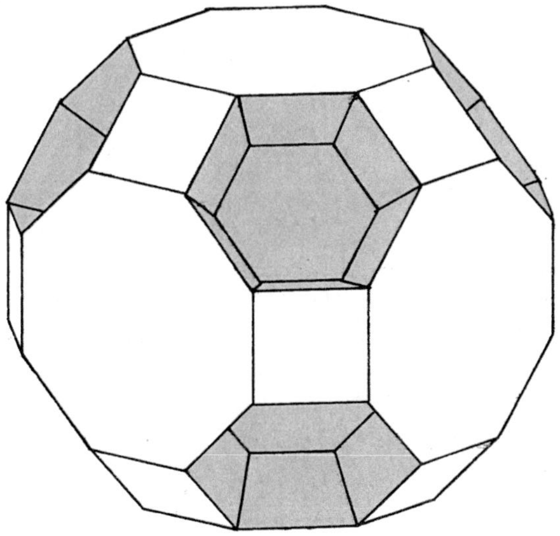

nt^2t^2{8,4 |3 2} [o{3,4,2}]
Truncated Cuboctaprism Octatetraskew

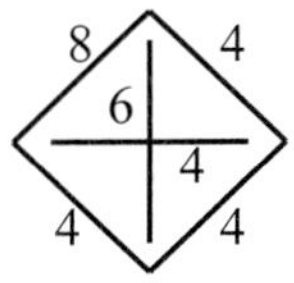

Fig.68: Octatetraskews

MORE OCTATETRASKEWS

Solids	V	E	F	S
octagon	48	96	12 ⯃	6
tr cuboctahedron			24 □	2
square			(8) ⬡	12
			(24) ‖	
Truncated Cuboctahedral				
square	24	72	12 □	6
rhombicuboctahedron			24 □	2
square			(8) △	12
			(24) ‖	
Rhombicuboctahedral				
cube	48	120	12 □	6
rhombicuboctahedron			48 □	2
triprism			(16) △	8
cube			(24) □	12
Rhombicuboctaprism				

Table 46: More Octatetraskews

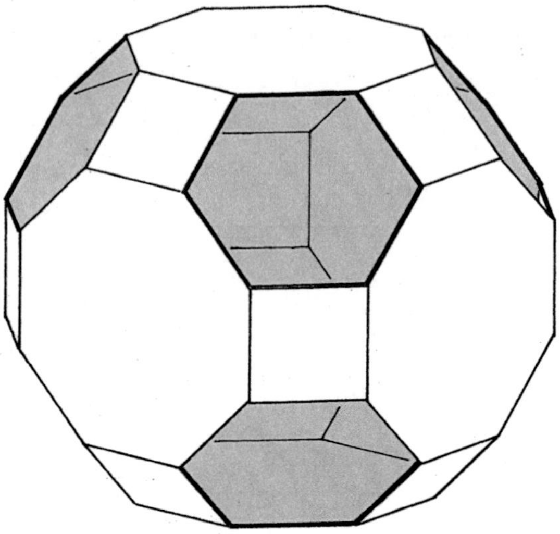

$[tt^2\{3,4,2\}]$

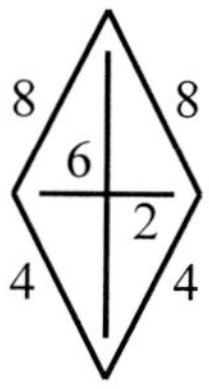

Truncated
Cuboctahedral
Octatetraskew

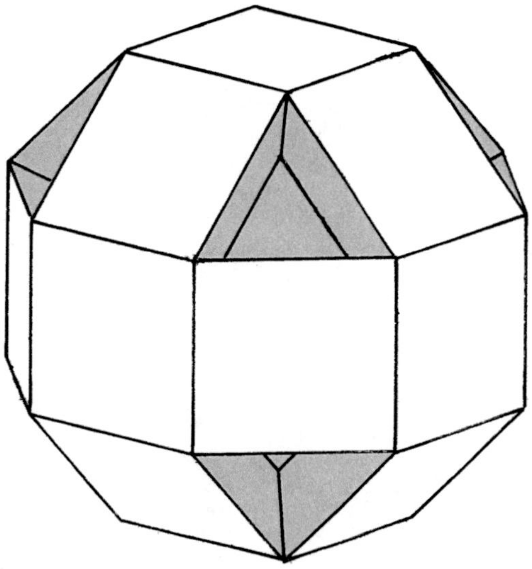

$[o^2\{3,4,2\}]$

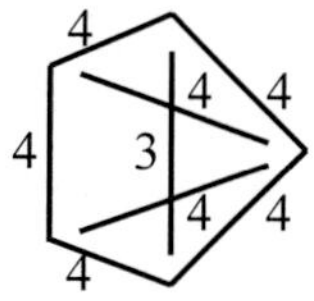

Rhombicuboctaprism
Octatetraskew

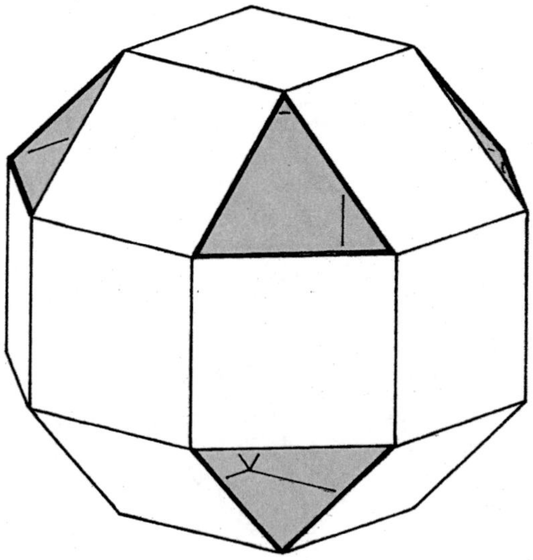

$[t^2t^2\{3,4,2\}]$

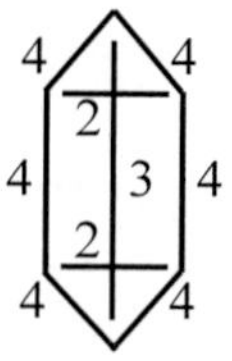

Rhombicuboctahedral
Octatetraskew

Fig.69: More Octatetraskews

EVEN MORE OCTATETRASKEWS

Solids	V	E	F	S
octaprism	48	96	12 ⯃	6
truncated cube			24 □	2
triprism			(16) △	8
			(12) □	
Truncated Cuboprism				
cube	24	72	12 □	6
cuboctahedron			24 □	2
triprism			(16) △	8
			(12) ‖	
Cuboctaprism				
cube	48	120	12 □	6
tr octahedron			48 □	2
hexaprism			(16) ⬡	8
square			(24) ‖	12
Truncated Octaprism				

Table 47: Even More Octatetraskews

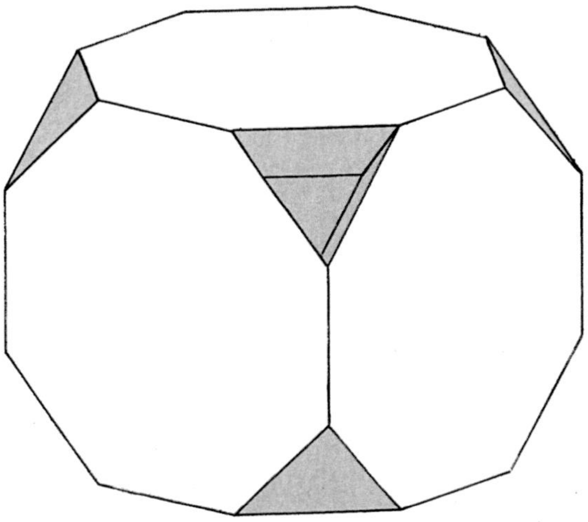

[$tt^2\{2,4,3\}$]

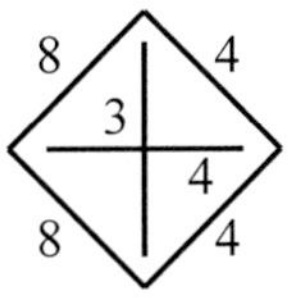

Truncated Cuboprism Octatetraskew

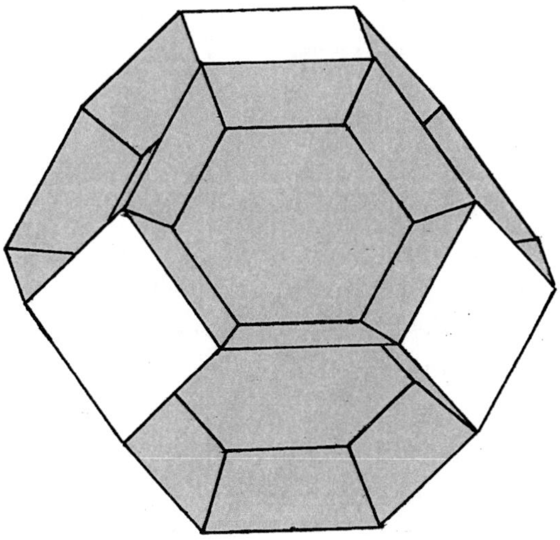

[$o^2\{2,4,3\}$]

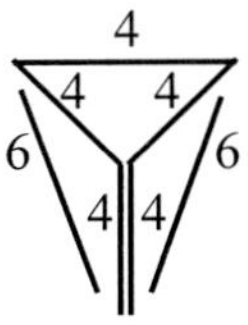

Truncated Octaprism Octatetraskew

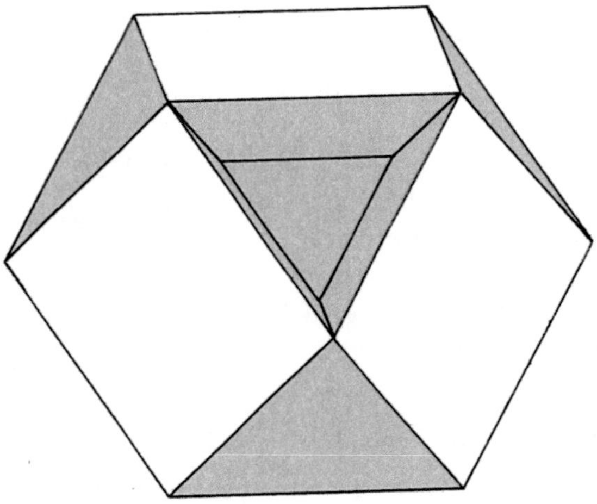

[$t^2t^2\{2,4,3\}$]

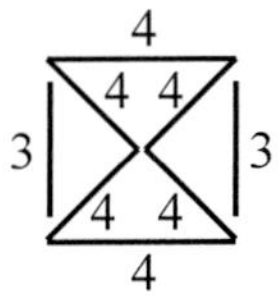

Cuboctaprism Octatetraskew

Fig.70: Even More Octatetraskews

DODECATETRASKEW GRID

Finally taking the third possible alignment of the central tetrahedral vertex figure of o{4,3,2}, we can miss out the two square faces of the truncated cuboctahedra, making them 12-connected.

These link to 2-connected cubes to form one space, whilst the octagon prisms, with some missing square faces, are 4-connected to form the other space linked by 3-connected hexagon prisms.

We have accordingly called these 'dodecatetraskews' and again this vertex can be transformed to produce another full grid of nine members as shown adjoining in Figure 71.

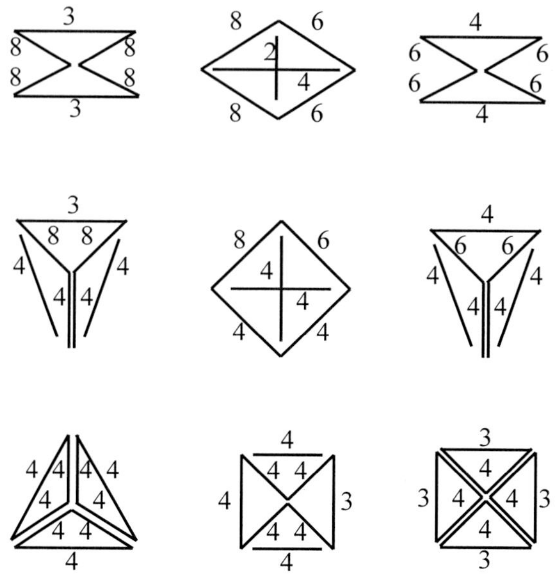

Fig.71: Dodecatetraskew Grid [3-4,2-12]

All of these are again topologically similar surfaces between a 12-connected space and a 4-connected one. Taking double vertices and edges into account, they also have a consistent Euler Number (V-E+F) of -20, again the sum of the Euler Numbers of the previous two families.

Two of the resultants in this grid are also based on $t^3\{4,3,2\} = t^3\{2,3,4\}$ and $o^3\{4,3,2\} = o^3\{2,3,4\}$ respectively, and along with the central member are shown opposite in Figure 72, with Table 48 describing their numerical quantities.

Figure 73 and Table 49 on page 78 show the top left members of this grid, two intermediates based on the $o^2\{2,3,4\}$ and $tt^2\{4,3,2\}$, and the resultant in the corner based on t{4,3,2}.

Similarly Figure 74 and Table 50 on page 79 illustrate the bottom right members of this grid, with its two intermediate forms based on $tt^2\{2,3,4\}$ and $o^2\{4,3,2\}$, and the resultant in the corner based on t{2,3,4}.

As with the previous set of families with fivefold symmetry, these with fourfold symmetry have a similar distribution of members where sleeves and slots appear as the connections.

The Hexatriskews and Octatetraskews have but one each, {4,6 |4 2} and {4,8 |3 2}, whilst the Dodecatetraskews on the following pages have four with sleeves and two with slots.

DODECATETRASKEWS

Solids	V	E	F	S
hexagon	24	72	12 □	8
tr octahedron			16 ⬡	2
square			(24) ‖	6
digon				12
Truncated Octahedral				
hexaprism	96	192	12 ⯃	8
tr cuboctahedron			16 ⬡	2
octaprism			48 □	6
cube			(24) □	12
			(24) □	
Truncated Cuboctaprism				
3-lines	16	72	12 □	8
cube			24 □	2
cube			(24) ‖	6
square			(24) ‖	12
Cuboprism				

Table 48: Dodecatetraskews

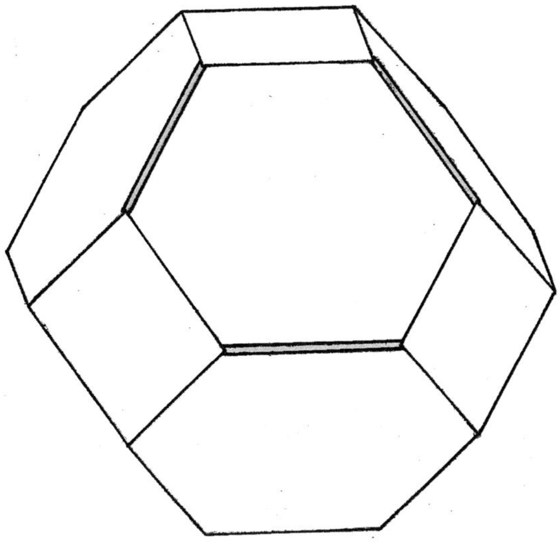

[t^3 {4,3,2}]
[t{3,4,2}]

Truncated
Octahedral
Dodecatetraskew

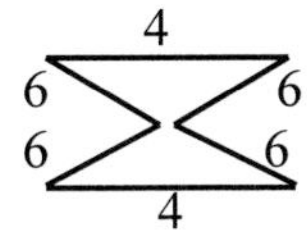

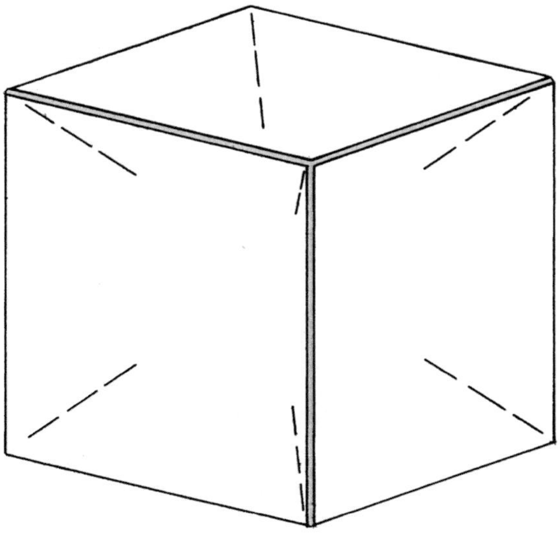

[o^3 {4,3,2}]
[t{2,4,3}]

Cuboprism
Dodecatetraskew

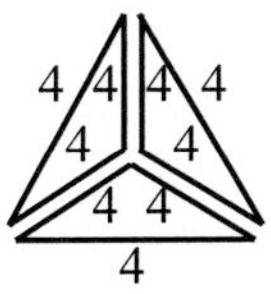

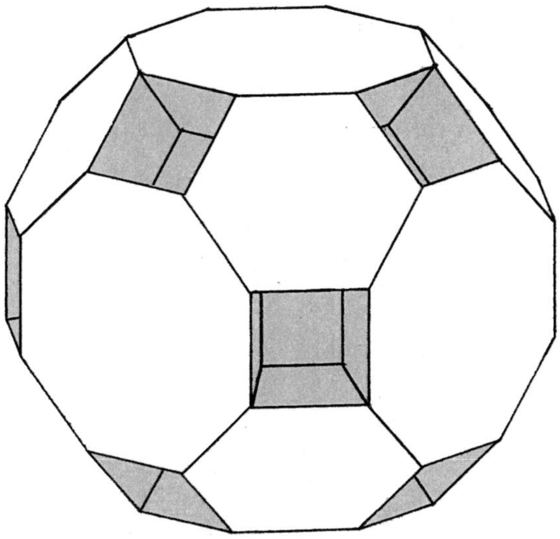

[o{4,3,2}]

Truncated
Cuboctaprism
Dodecatetraskew

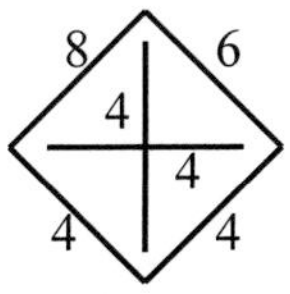

Fig.72: Dodecatetraskews

MORE DODECATETRASKEWS

Solids	V	E	F	S
hexagon	48	96	12 ⯃	8
tr cuboctahedron			16 ⬡	2
octagon			(12) □	6
square			(24) ‖	12
Truncated Cuboctahedral				
triangle	24	72	12 ⯃	8
truncated cube			16 △	2
octagon			(24) ‖	6
digon				12
Truncated Cubic				
triprism	48	144	12 ⯃	8
truncated cube			16 △	2
octaprism			48 □	6
square		(24) ‖	(24) □	12
Truncated Cuboprism				

Table 49: More Dodecatetraskews

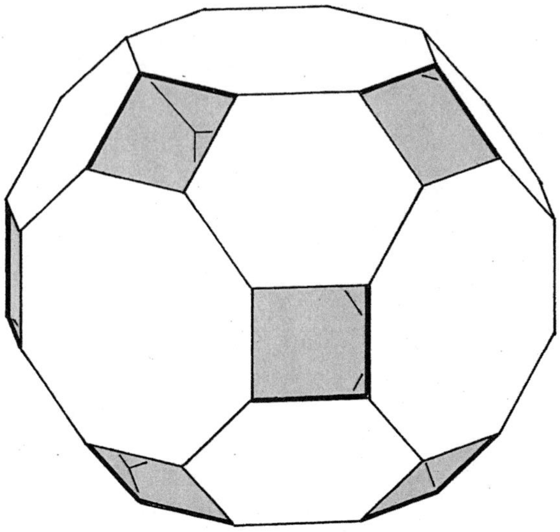

Truncated Cuboctahedral Dodecatetraskew

[tt^2{4,3,2}]
[tt^2{3,4,2}]

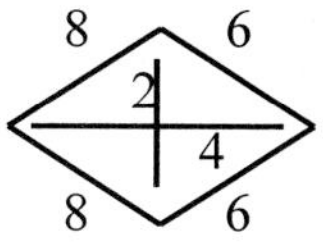

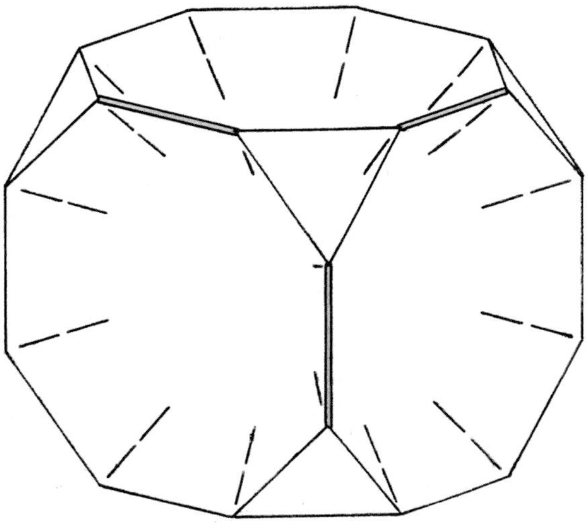

Truncated Cuboprism Dodecatetraskew

[o^2{2,3,4}]
[tt^2{2,4,3}]

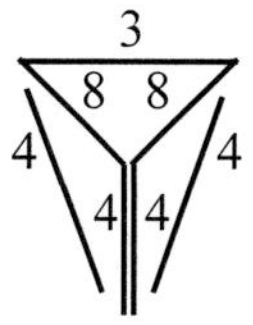

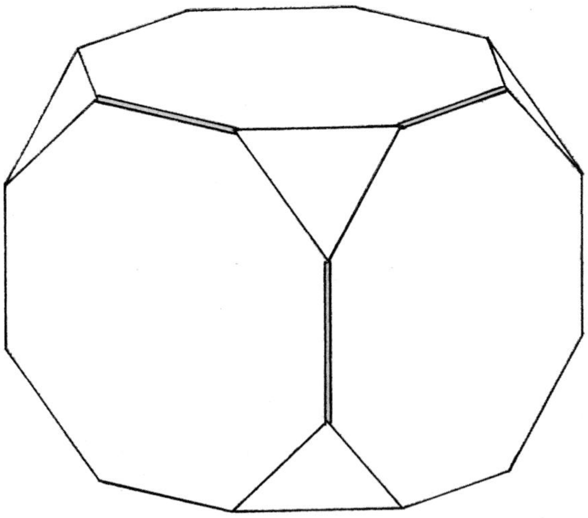

Truncated Cubic Dodecatetraskew

[t^3{3,4,2}]
[t{4,3,2}]

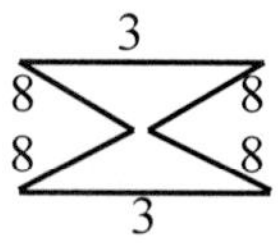

Fig.73: More Dodecatetraskews

EVEN MORE DODECATETRASKEWS

Solids	V	E	F	S
hexaprism	48	144	12 □	8
tr octahedron			16 ⬡	2
cube			48 □	6
square		(24) ∥	(24) □	12
Truncated Octaprism				
triprism	12	72	16 △	8
octahedron			24 □	2
4-lines			(24) ∥	6
square			(24) ∥	12
Octaprism				
triprism	48	144	12 □	8
rhombicubocta			16 △	2
cube			48 □	6
cube		(24) ∥	(24) □	12
Rhombicuboctaprism				

Table 50: Even More Dodecatetraskews

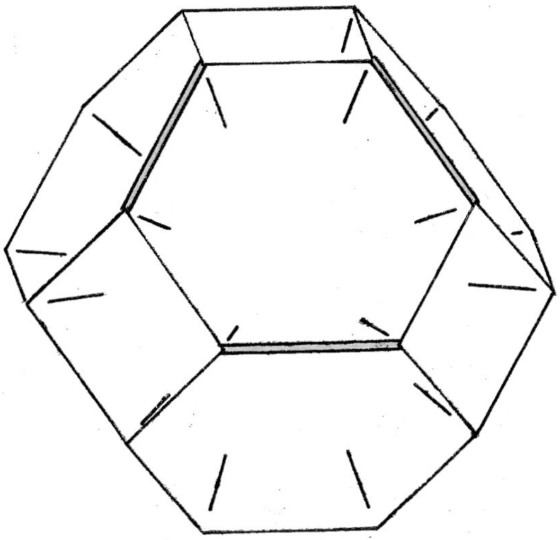

[tt^2{2,3,4}]
[o^2{2,4,3}]

Truncated Octaprism Dodecatetraskew

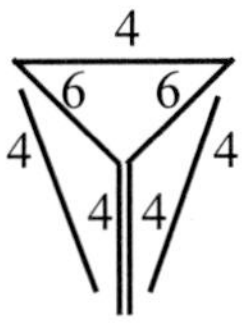

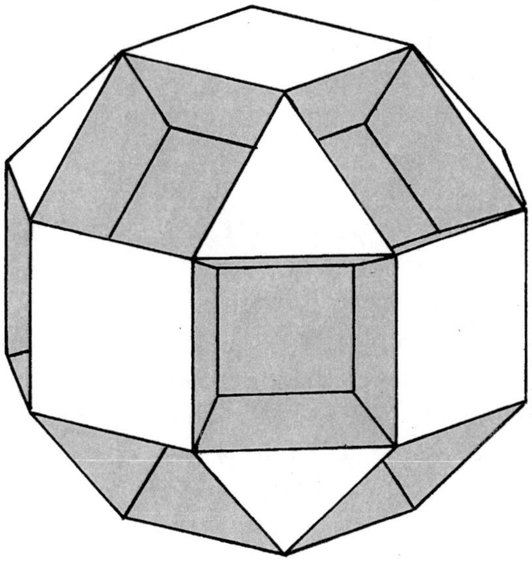

[o^2{4,3,2}]
[o^2{3,4,2}]

Rhombicuboctaprism Dodecatetraskew

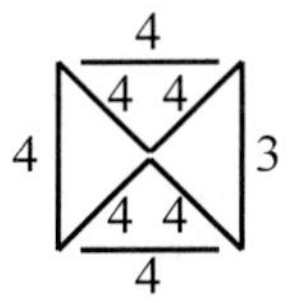

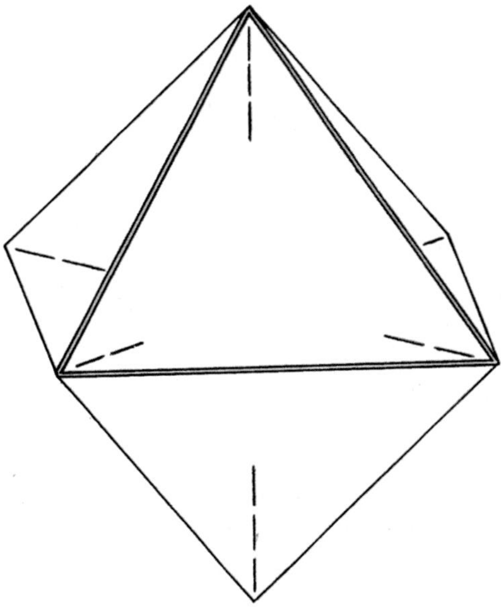

[o^3{3,4,2}]
[t{2,3,4}]

Octaprism Dodecatetraskew

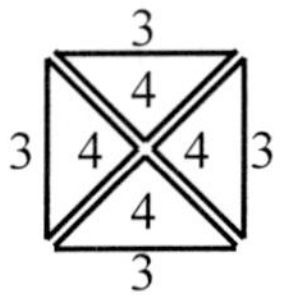

Fig.74: Even More Dodecatetraskews

TETRATRISKEW GRID

Finally we come to consider the tetrahedron as a four dimensional polytope, {3,3,2}, comprising two tetrahedral cells {3,3}, two per edge. These two tetrahedra back to back are again a higher dimensional version of the dihedra we examined following the polyhedra in the appendix to *'The Simpler? Polyhedra'*.

In four dimensions, viewed from three, there appear to be two tetrahedra, an inner and an outer one that share all their faces, a direct analogy to the dihedra that are back to back polygons.

The grid for this family, shown adjoining in Figure 75, has a central member based on the first odd, o{3,3,2} and two parents top right and bottom left based on t^3{3,3,2} and o^3{3,3,2}] respectively. These three are shown opposite in Figure 76, and their numerical properties are set out in Table 51.

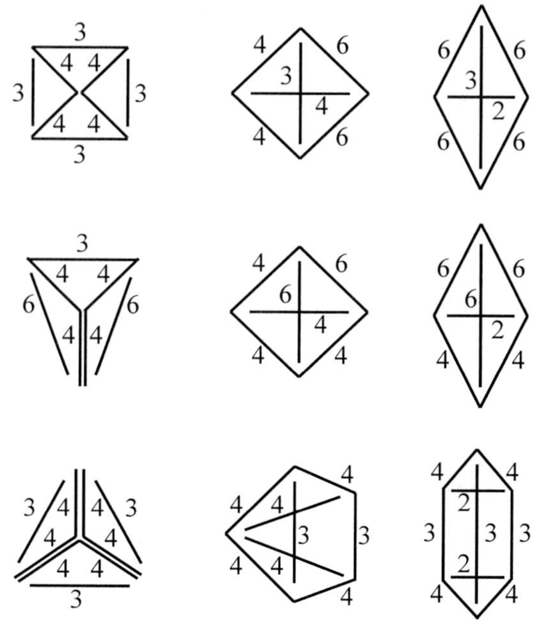

Fig.75: Tetratriskew Grid [2-3,2-4]

The truncated tetrahedral tetratriskew {6,4 |3 2} comprises only 8 hexagons in back to back pairs, four per vertex and 3-connected to each other around the sphere via missing digons. The two truncated tetrahedra, one inner one outer, are 4-connected to each other via the missing triangle faces.

Its dual, the tetraprism tetratriskew {4,6 |3 2} exhibits the same connectivities. Each 'three lines' is 3-connected to paired back to back squares (digon prisms as sleeves) linking around the sphere, whilst the hollow triangle prisms between the inner and outer tetrahedra provide the necessary 4-connections.

The truncated octaprism tetratriskew, central to this grid, is similarly connected as can more readily be seen in the illustration.

It has half its hexaprisms 3-connected to each other by cubes and its 4-connected truncated octahedra linked by the other hexaprisms.

As before all members of this family grid share a common Euler Number of -4. The remainder of this family also share these connectivities and are shown overleaf. Figure 77 shows the two intermediates and a resultant from bottom right of the grid, as detailed in the adjoining Table 52, whilst Figure 78 shows those at top left, as detailed in Table 53.

Because the tetrahedron is self-dual the next alignment of the central tetrahedral vertex here will only produce the same grid of possibilities. The next section after this grid's members have been described will thus present the third alignment (page 84).

TETRATRISKEWS

Solids	V	E	F	S
hexagon	12	24	8 ⬡	4
tr tetrahedron			(4) △	2
{6,4 |3 2}			(6) ‖	
Truncated Tetrahedral				
hexaprism	48	96	8 ⬡	4
tr octahedron			36 □	2
hexaprism			(8) ⬡	4
cube			(12) □	6
nt^2t^2{4,6 |3 2}				
Truncated Octaprism				
3-lines	8	24	12 □	4
cube			(8) △	2
cube			(12) ‖	4
square				6
{4,6 |3 2}				
Tetraprism				

Table 51: Tetratriskews

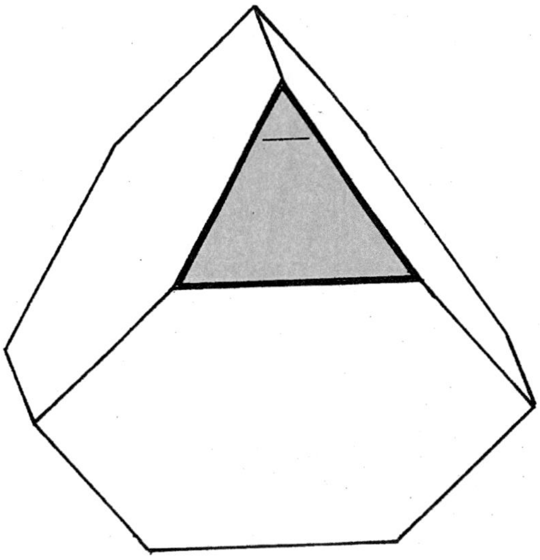

{6,4 |3 2} Truncated Tetrahedral Tetratriskew

$[t^3\{3,3,2\}]$
$[t\{3,3,2\}]$

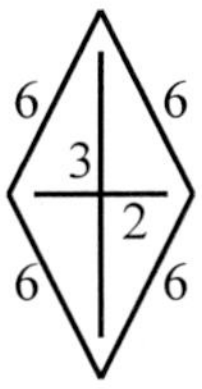

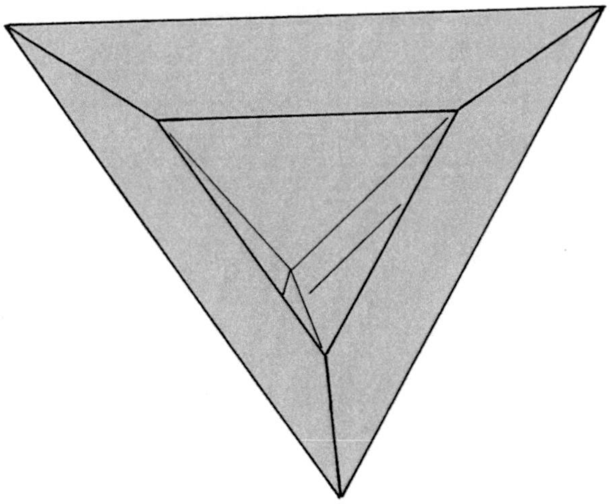

{4,6 |3 2} Inverted Tetraprism Tetratriskew

$[o^3\{3,3,2\}]$
$[t\{2,3,3\}]$

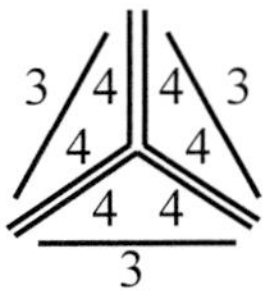

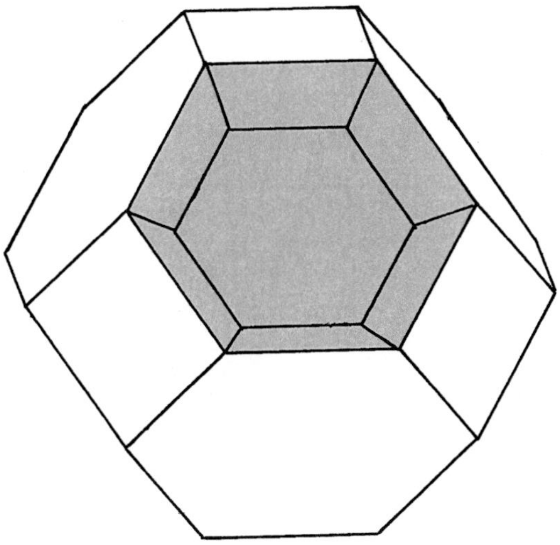

nt^2t^2{4,6 |3 2} Truncated Octaprism Tetratriskew

$[o\{3,3,2\}]$

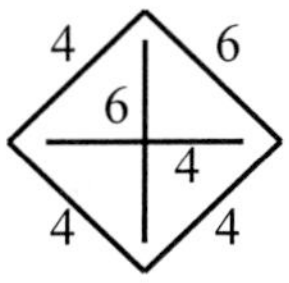

Fig.76: Tetratriskews

Solids	V	E	F	S
hexagon	24	48	8 ⬡	4
tr octahedron			12 □	2
square			(4) ⬡	6
			(12) ∥	
Truncated Octahedral				
triangle	12	36	8 △	4
cuboctahedron			12 □	2
square			(4) △	6
			(12) ∥	
Cuboctahedral				
triprism	24	60	8 △	4
cuboctahedron			24 □	2
triprism			(8) △	4
cube			(12) □	6
Cuboctaprism				

Table 52: More Tetratriskews

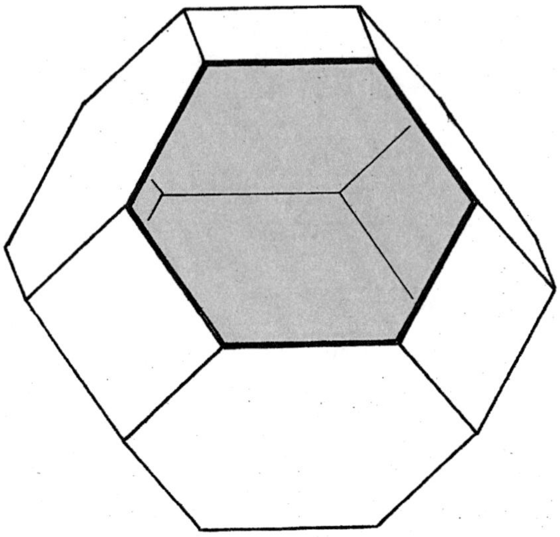

$[tt^2\{3,3,2\}]$

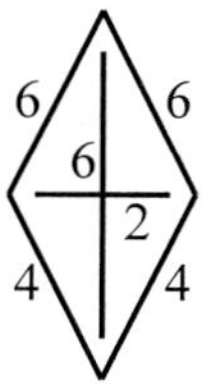

Truncated Octahedral Tetratriskew

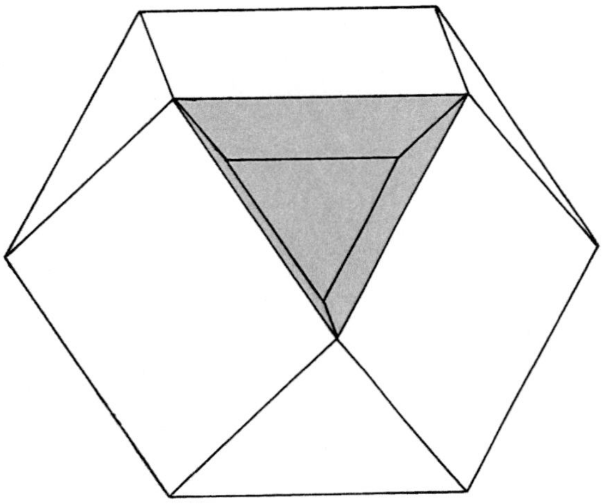

$[o^2\{3,3,2\}]$

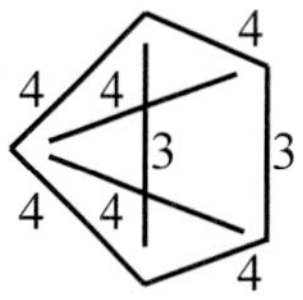

Cuboctaprism Tetratriskew

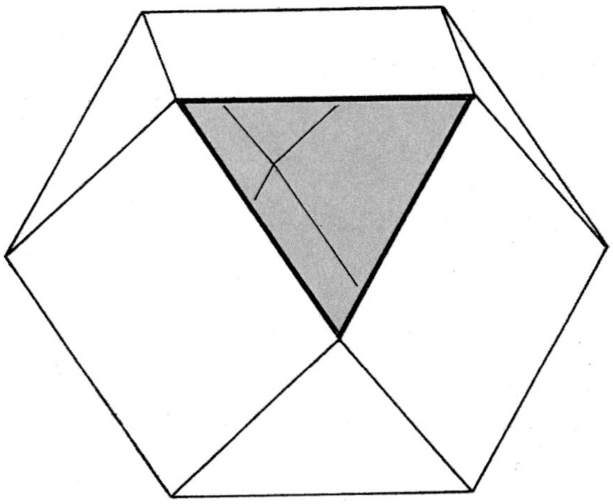

$[t^2t^2\{3,3,2\}]$

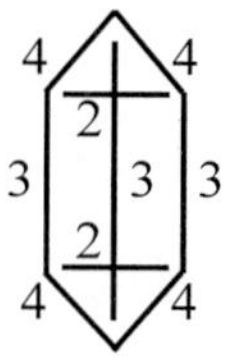

Cuboctahedral Tetratriskew

Fig.77: More Tetratriskews

EVEN MORE TETRATRISKEWS

Solids	V	E	F	S
hexaprism	24	48	8 ⬡	4
tr tetrahedron			12 □	2
triprism			(8) △	4
			(6) □	
Truncated Tetraprism				
triprism	12	36	8 △	4
octahedron			12 □	2
triprism			(8) △	4
			(6) ‖	
Octaprism				
triprism	24	60	8 △	4
tr tetrahedron			24 □	2
hexaprism			(8) ⬡	4
square			(12) ‖	6
Truncated Tetraprism				

Table 53: Even More Tetratriskews

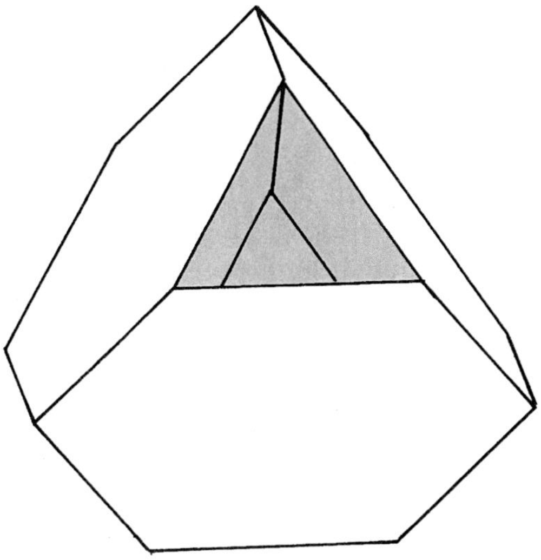

$[tt^2\{2,3,3\}]$

Truncated Tetraprism Tetratriskew

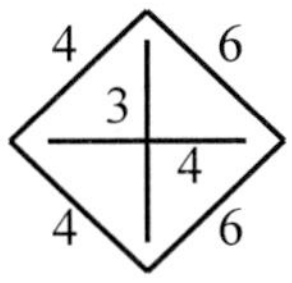

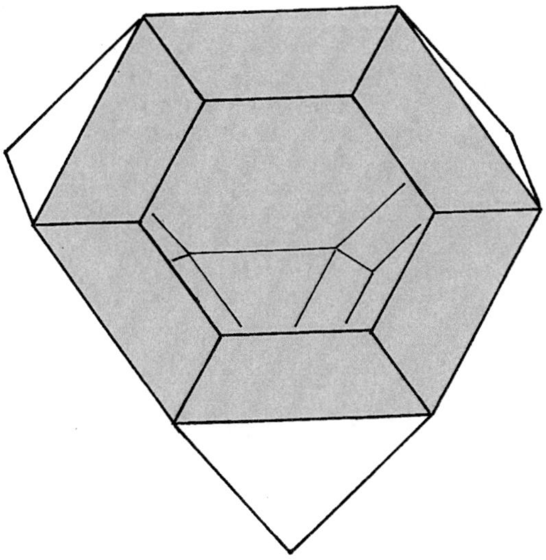

$[o^2\{2,3,3\}]$

Inverted Truncated Tetraprism Tetratriskew

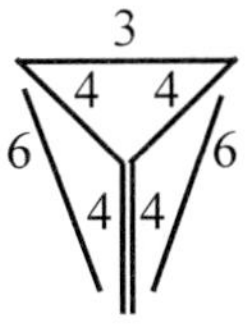

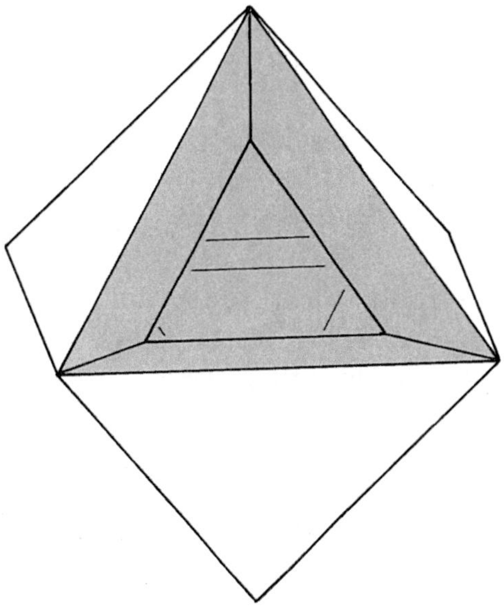

$[t^2t^2\{2,3,3\}]$

Octaprism Tetratriskew

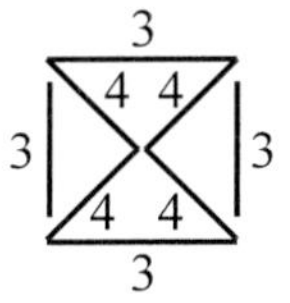

Fig.78: Even More Tetratriskews

HEXATRISKEW GRID

Finally taking the third possible alignment of the central tetrahedral vertex figure of o{3,3,2}, we can miss out the two square faces of the truncated octahedra, making them 6-connected.

These link to 2-connected cubes to form one space, whilst the hexagon prisms, with some missing square faces, are 3-connected to form the other space linked by further 3-connected hexagon prisms.

We have thus called these 'hexatriskews' and again this vertex can be transformed to produce another full grid of nine members as shown adjoining in Figure 79.

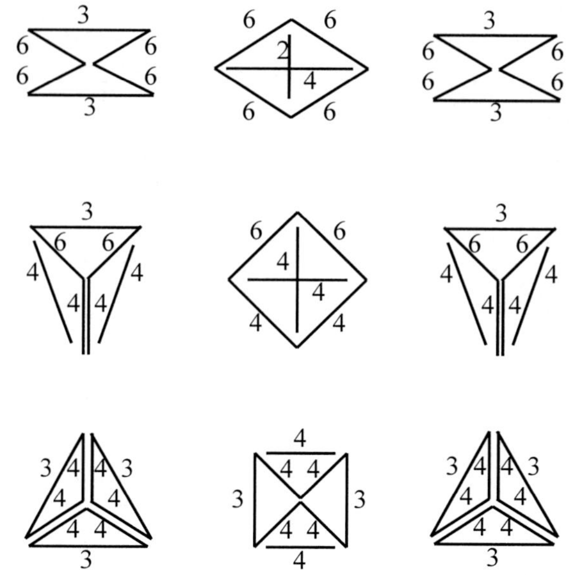

Fig.79: Hexatriskew Grid [3-3,2-6]

All of these are again topologically similar surfaces between a 6-connected space and a 3-connected one. Taking double vertices and edges into account, they also have a consistent Euler Number (V-E+F) of -8, again the sum of the Euler Numbers of the previous two (identical) families (both -4).

The three members of the central column of this particular grid have already been seen as the top row of a previous family's grid of hexatriskews [2-3,2-6] (page 68). It can also be seen that the two outer columns are also identical, so that we only have three further examples to illustrate.

These are shown adjoining in Figure 80 and described numerically in Table 54. We have thus now completed our examination of all five platonic solids as generators in four dimensions of degenerate skew polyhedra.

HEXATRISKEWS

Solids	V	E	F	S
hexagon	12	36	8 △	4
tr tetrahedron			8 ⬡	2
triangle			(12) ‖	4
digon				6
Truncated Tetrahedral				
hexaprism	24	72	8 △	4
tr tetrahedron			8 ⬡	2
triprism			24 □	4
square		(12) ‖	(12) □	6
Truncated Tetraprism				
3-lines	8	36	8 △	4
tetrahedron			12 □	2
triprism			(12) ‖	4
square			(12) ‖	6
Tetraprism				

Table 54: Hexatriskews

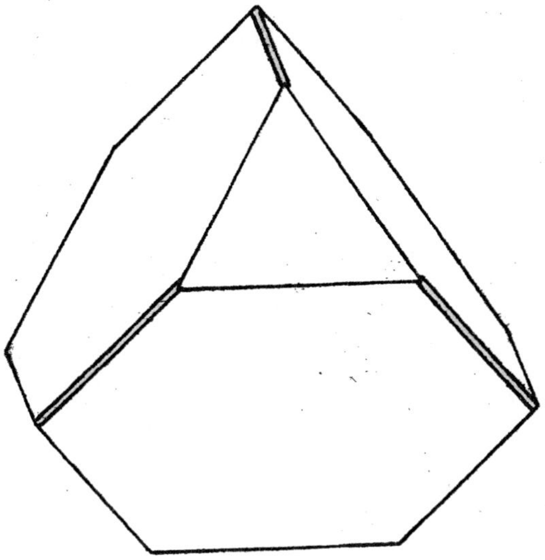

Truncated Tetrahedral Hexatriskew

[t^3{3,3,2}]
[t{3,3,2}]

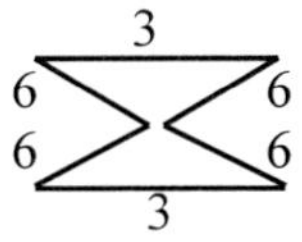

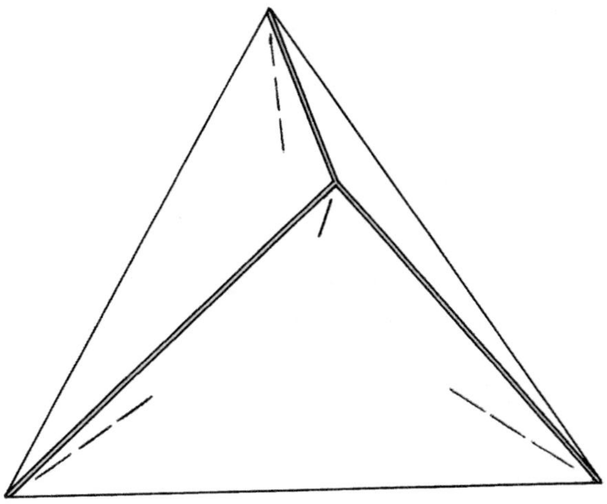

Tetraprism Hexatriskew

[o^3{3,3,2}]
[t{2,3,3}]

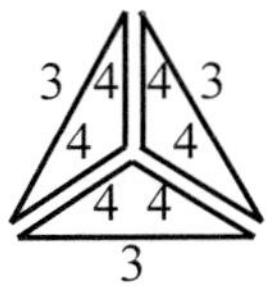

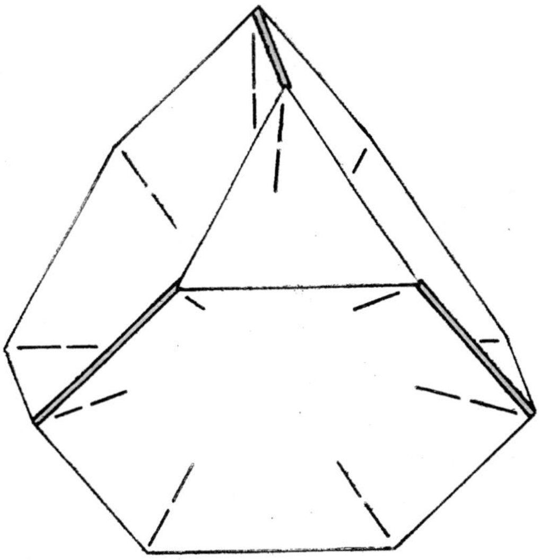

Truncated Tetraprism Hexatriskew

[tt^2{2,3,3}]

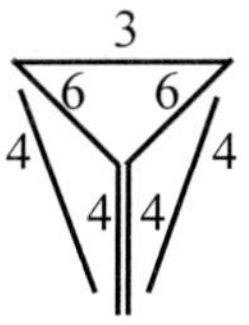

Fig.80: Hexatriskews

DEGENERATES SUMMARY

We saw in the last volume of this work, *'The Skew Polyhedra',* that the degenerate skews based around the tilings {6,3}, {3,6} and {4,4} were produced by considering those tilings as four dimensional forms belonging to three families {6,3,2}-{2,3,6}, {3,6,2}-{2,6,3} and {4,4,2}-{2,4,4} in turn.

As with the skew polyhedra themselves, each family's first odd member produced three grids of nine members each. Again within each grid the connectivities were preserved despite the differing polygon faces involved.

The immediately preceding pages here have now applied those same principles to the polyhedra considered as four dimensional forms. This produced the degenerate hyperskews, the folded up versions of the tiling degenerates.

These were based on the polyhedra {5,3}, {3,5}, {4,3}, {3,4} and {3,3} in turn, to produce the four dimensional analogues of the degenerate skew polyhedra. These again preserve connectivities within a grid and essentially comprise forms where two equal polyhedra form the faces of four dimensional prisms and have connecting holes through them.

The inherent beauty of these is that what we see in a three dimensional projection as an outer and an inner polyhedron connected by prisms, are actually in four dimensions two exactly equal polyhedra. The enhanced symmetry of the torus when seen in four dimensions (p.52) comes to mind here. As does the hypercube with its inner and outer cubes conjoined by what we see as six truncated pyramids that are actually cubes too.

HYPERSKEWS SUMMARY

Overall then we have successfully applied the principle of vertex transformation of tetrahedral vertex figures to a wide range of possibilities.

In the last volume we first found the usual skew polyhedra that comprise surfaces stretching to infinity in three dimensions in and around the closepacks and their truncations. These were then augmented by the degenerate skew polyhedra that stretched to infinity in two dimensions in and around the tilings and their truncations, considered as four dimensional objects.

Here in this volume we first considered the hyperskew polyhedra that comprise finite surfaces stretching in and around the hypersolids and their truncations. These all had characteristic connectivities between adjacent cells and constant Euler numbers within any particular family represented by a nine member grid of vertex figures.

These in their turn were augmented by the degenerate hyperskew polyhedra, which are also finite surfaces, but wrapping in and around the polyhedra considered as four dimensional objects, essentially folded up tilings of the sphere. These too had characteristic connectivities between adjacent cells and constant Euler numbers within any particular family represented by a nine member grid of vertex figures.

So we must ask now is there any more? Well there is one small area we need to look at in a further volume. This will concern similar surfaces related to both the skew polyhedra and the hyperskews, where the connections through between adjacent cells are constricted to mere points, leading us on to a consideration of 'stick models'.

BIBLIOGRAPHY

Abbott, E A 1962
Flatland
Blackwell

Aste, T & Weaire, D 2000
The Pursuit of Perfect Packing
Institute of Physics

Coxeter, H S M 1937
Regular Skew Polyhedra
Proc. London Mathematical Soc. p33

Coxeter, Longuet-Higgins & Miller 1953
Uniform Polyhedra
Phil. Trans. R. Soc. Lond. A246, p401

Coxeter, H S M 1963
Regular Polytopes
Macmillan

Coxeter, H S M 1991
Regular Complex Polytopes
Cambridge UP

Critchlow, K 1969
Order in Space
Thames & Hudson

Cromwell, P R 1997
Polyhedra
Cambridge UP

Grünbaum, B & Shephard, G C 1989
Tilings and Patterns: An Introduction
Freeman

Holden, A 1971
Shapes, Space and Symmetry
Columbia UP

Kappraff, J 1990
Connections
McGraw-Hill

Skilling, J 1974
The Complete Set of Uniform Polyhedra
Phil. Trans. R. Soc. Lond. A278, p111

Taylor, P 1997
The Complete? Polygon
Nattygrafix

Taylor, P 1998
Incomplete Tilings
Nattygrafix

Taylor, P 1999
The Simpler? Polyhedra
Nattygrafix

Taylor, P 2000
The Star & Cross Polyhedra
Nattygrafix

Taylor, P 2004
Closepacks & Quasi-Closepacks
Nattygrafix

Taylor, P 2006
Hypersolids & Quasi-Hypersolids
Nattygrafix

Taylor, P 2013
The Skew Polyhedra
Polystar Press

Wenninger, M J 1971
Polyhedron Models
Cambridge UP

Weyl, H 1952
Symmetry
Princeton UP

Williams, R 1972
Natural Structure
Constable

OF RELATED INTEREST:

THE SIMPLER? POLYHEDRA
by Patrick Taylor 1999

being the third part of several comprising the complete? polyhedra

presents the simpler polyhedra derivable from the platonic solids; their truncations, quasitruncations, snub forms and the simpler sub-families and hemihedra, along with an appendix describing the prismatic dihedral forms.

80pp A4 paperback including
62 line drawings and 29 tables

ISBN 0 9516701 4 X £6.00

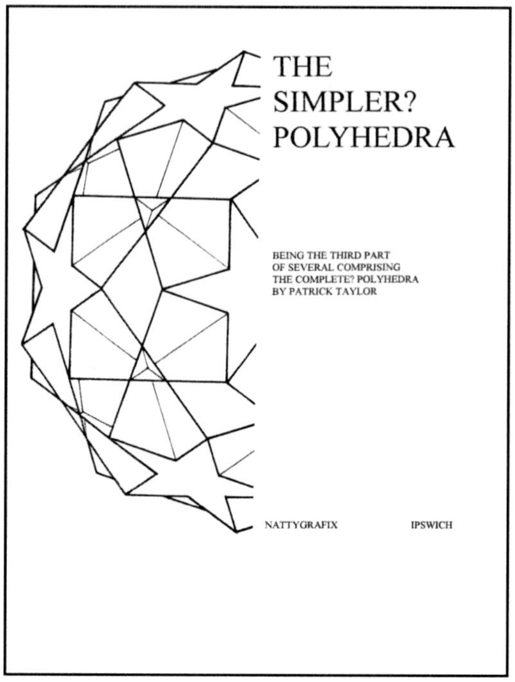

HYPERSOLIDS & QUASI-HYPERSOLIDS
by Patrick Taylor 2006

being the sixth part of several comprising the complete? polyhedra

presents the hypersolids as assemblies of polyhedra that inhabit four dimensions; includes truncations and quasitruncations; investigating fully the {3,3,3}, (3,4,3}, {3,3,4}-{4,3,3} and {3,3,5}-{5,3,3} symmetry families

80pp A4 paperback including over
54 line drawings and 44 tables

ISBN 0 9516701 7 4 £6.00

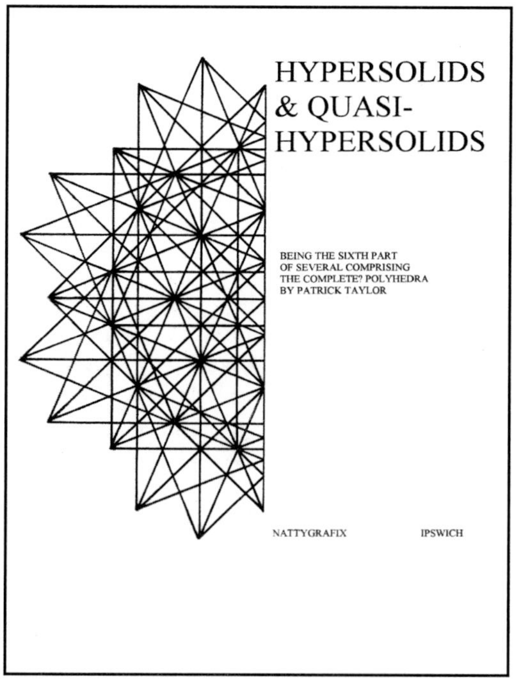